KNOWLEDGE AND IGNORANCE

Essays on Lights and Shadows

Folke Dovring

Westport, Connecticut
London

Library of Congress Cataloging-in-Publication Data

Dovring, Folke.
    Knowledge and ignorance : essays on lights and shadows / Folke
Dovring.
        p.   cm.
    Includes bibliographical references and index.
    ISBN 0–275–96139–7 (alk. paper)
    1. Science—Philosophy.   2. Knowledge, Theory of.   I. Title.
Q175.D699   1998
501—dc21   97–33244

British Library Cataloguing in Publication Data is available.

Library of Congress Catalog Card Number: 97–33244
ISBN: 0–275–96139–7

First published in 1998

Praeger Publishers, 88 Post Road West, Westport, CT 06881
An imprint of Greenwood Publishing Group, Inc.

Printed in the United States of America

The paper used in this book complies with the
Permanent Paper Standard issued by the National
Information Standards Organization (Z39.48–1984).

10 9 8 7 6 5 4 3 2 1

Contents

Preface

More than half a century of research and writing has given me ample time to reflect on the limits of our intellectual endeavors. My lifetime of search and thinking has led me traveling across countries and continents and civilizations, as well as across disciplines and problem areas.

Growing up in the family of a famous poet in the middle of old-fashioned peasant country (Sweden), I experienced nature more intensely than do most modern people. The vagaries of family history got me started as a historian of the Middle Ages. Breaking this mold, I first veered over into economic history and then, by way of U.N. service, to agricultural economics. This environment let my early experience of nature come to renewed life. The total exposure of intellectual cross-currents has been breathtaking at times.

Along the way I have been privileged to converse with many scholars with specialties as diverse as linguistics, soil science, legal concepts, and energy physics. I owe intellectual debts to many of them, but none has been directly engaged with the preparations for this book. What errors it may contain are all mine.

During the whole journey of life, the problems of knowledge and ignorance have followed me like a shadow. In formulating, at long last, the conclusions of my labors on these weighty matters, I feel humbled by the power of the subject itself. To give some proportion to my endeavor, let me cite a few lines by a scholar of towering stature whom it was my fate to encounter in a way that forced me to toil over a text I happened to discover. Its author, Hugo Grotius, in an apology for one of his own works, wrote (and I paraphrase rather than translate precisely): The perfection of any matter cannot be gauged by abstract speculation, but only by what the matter at hand may require. In other words: Judge me not by any impossible ideal, only by what I may have contributed to the debates that must go on.

ONE

Beyond Science:
Limits and Problems

There are more things in heaven and earth . . . Than are dreamt of in
your philosophy

Shakespeare, *Hamlet*, Act 1, scene 5: 164–65

As science keeps enlarging its domain by discoveries that were undreamt
of not long ago, its borders against the still unknown become longer and
less well-defined. None of this would be problematic except for the continued craving of the human spirit for certainties that science is unable
to supply. The acrimony surrounding evolution versus creation is symptomatic, but it is only one among many similar conflicts, most of them
less well understood, in which basic values are at stake. This means,
among other things, that we need to define science and its basic concepts
in ways that the lay mind can comprehend. This will also redefine some
of the areas where science should not tread. We need to realize that the
lights of knowledge tend to define their own shadows of ignorance.

Early science became the victim of its own method because it could
see none other. "Man is a machine," wrote LaMettrie in 1748, for his
scientific technique could discover only the mechanical elements in our
nature. The Age of Reason had just witnessed a major outbreak of crude
superstition in the witch persecutions that only cold reason seemed able
to overcome.

From early mechanistic thinking, the line goes on through Pavlov, to Communist "dialectical materialism," and to recent American behaviorists. They believe what they can see and deny what they cannot see. The limits of sensory perception were only beginning to be acknowledged in the idealism of Berkeley and Hume. Philosophical materialism still took center stage in nineteenth-century science, well before matter began to be relativized by nuclear physics. Ludwig Büchner's "Energy and Matter" (*Kraft und Stoff*, 1855) may be crude, but it is typical of the intellectual trap scientism set for itself. The line continues through Haeckel and Ostwald and far beyond. Latter-day materialism is sometimes more subtle, but still seeks to maintain its monopoly on truth by insisting on a material base as the essence of all reality.

For many modern people, science stands as the only valid source of light—of insight into reality. To maintain that science can answer all legitimate questions about reality is a logical leap that may be understandable at a certain stage of intellectual development. It is nonetheless a fallacy and will not only mislead, but will also provoke legitimate opposition. Not all believers in scientism are honest enough to admit that their faith in science is just an article of faith, not a scientific statement.

So many things have changed in the world of our knowledge since the days of the early materialists that it should be time—it is long overdue, in fact—to call one-sided scientism to account for its arrogance as the sole source of intellectual power. It has dawned on many of us that science is, to a large extent, a set of approximations as cover for engineering. Absolute truth is something else. Even mathematics has not escaped entirely from this kind of problem.

Some revision of basic concepts has begun among the bolder spirits in modern physics, starting with Niels Bohr and subsequently spearheaded by Ilya Prigogine and Hannes Alfvén and some more recent writers (Collins and Finch 1993). There are also many tentative departures toward holistic thinking, as in some new styles of medical care.

The next "scientific revolution" must focus on the limits of our current knowledge, the limits of our instruments of knowledge, and the awesomely vast scope of the still unknown. Some of our overwhelming ignorance we may have to live with forever. Sooner or later we shall have to come to terms with this basic source of conflict. We shall also have to come to terms with the value aspect of science: What knowledge is more worth having—about the gates of Hell or the meadows of Paradise?

None of this, of course, gives any degree of justification for the so-called "creation science." Absurd as it is, Bible-based creationism, by its very existence, demonstrates the need for the scientific community to become more aware of absurdities originating from within itself.

The upheavals in physics and chemistry that followed upon the discovery of radioactivity should have taught us an essential lesson. Before Becquerel, every scientist worth his salt—from Dalton and Lavoisier to Mendeleyev and Ostwald—was calmly convinced that the chemical elements were immutable, hence the ultimate source of reality's qualities. The alchemist assumption was rejected for lack of positive proof. Becquerel's earth-shattering discovery was initially sometimes "pooh-poohed" as due to some overlooked chemical reaction. Even the Curies are said to have, at first, reacted with disbelief toward the consequences of their own discoveries, which are now commonplace: that chemical elements can be created and destroyed by fission, fusion, and particle bombardment.

This chain of events in scientific discovery should have warned us that fringe phenomena, as well as easily overlooked trivia, may have more significance than we readily credit them with. In the case of radioactivity, nature set an intellectual trap for us because, if spontaneous radioactivity had been a major, and an obvious, occurrence on Earth, then the planet would not have been suitable for organic life. We would then not have been here in the first place. We should note the lesson. It has many parallels; many other intellectual traps are set for us by the practical circumstances of our existence. Respect for the unknown should make us hold back judgment more often than we do and should make us avoid jumping to unwarranted conclusions. The legitimate domain of science is what can be proved positively. Lack of proof may lead to acquittal in a court of justice, but in the halls of science it should merely give pause.

Such reflections are necessary here because in reexamining the intellectual substance of science, we must keep in mind that the type of conclusions offered by mechanistic schools of thought, such as materialism and behaviorism, include some fallacies such as "the unproven does not exist."

We should also think twice about the so-called "parsimony principle" (or "Occam's razor") in scientific inquiry. The preference for simple solutions is a guide to the conduct of research, but it does not tell us what we should believe in. Testing the simplest hypothesis first will be less expensive in those cases where the simplest hypothesis turns out to be confirmed. In many cases, it is not, and then the procedure is not particularly economical. If it is allowed to set our minds to rest on matters that are not yet fully investigated, the parsimony principle turns into a huge fallacy. Neither logic nor experience tells us that simple solutions are more likely to be true than complex ones. It may merely seem that way in the many cases where our information is inadequate. Nature is not simple and human nature is even less so. Modern science has discovered many things to be so complex that previous generations of scientists rejected the true answers by applying the parsimony principle. Recent examples include plate tectonics (in geology) and the origin of mitochondria (in biology). In both cases, overcautious "conservative" thinking tended to withhold essential resources from research that in the end turned out to be closest to reality, complex as it was. Occam's razor became circular logic.

Let us therefore keep the parsimony principle within its proper role: as a strategy for research, not as a guide for thinking about the unknown. The recent vogue of applying and accepting paradigms as working hypotheses on vast areas of knowledge risks very much transferring Occam's razor from a tool to a theory.

The quantum jump in scientific knowledge represented by nuclear physics is more than an object lesson. It has also given us a powerful intellectual tool that helps to break wide open the whole traditional construct of the world as consisting of matter and energy. Werner Heisenberg, in probing the outer limits of physical reality, found that he could not really say with any certainty what is out there at those outer limits. The full consequence of reducing both matter and energy to elemental light quanta (or photons) appears slow in coming, however. There are recent departures, but as yet not any new system.

The lesson learned in physics seems, in any event, to have remained among the natural sciences. The social and behavioral sciences apparently do little to explore the bases of their own work, to speak nothing of the humanities. Any reorientation of scientific work, which the new physics will force along, should have even greater effects on the scientific study of human affairs because this is where known existence reaches its highest degrees of complexity.

The essays in this book will first explore some of the limitations of our scientific ambitions and then sketch some of the positive opportunities on which scientific work has proceeded and can proceed.

We will attempt to show how most people—including most scientists—have had their vision of reality limited by specific instruments of knowledge to which they are confined most of the time, by nature or by cultural and intellectual history. Strive as we may to reduce those limitations, they will always be there. We cannot lay claim to having divine omniscience within our reach, but we can acknowledge the basic problem of intellectual limits. By aid of a simile of a "lighted room," we will discuss the contrast of light and shadow. As in a sunlit or moonlit landscape, the brightness of light somehow defines the dark of the unlit areas. Other motives for limitation are in the subjectivity of much of our direct experience that we have just inside ourselves. Foremost are pain and other emotions. The subjectivity of aesthetic and religious experience also belongs here. Other absolute obstacles against knowledge are raised by nature, typically in animal feelings and mentality, which to a large extent are systemically withheld from our insight. Parapsychology may also include some areas that are just not accessible to the normal and reasoning human mind.

Against the persistent simplistic and mechanistic view that has dominated much of recent science, we now have powerful special tools. Among these are chaos theory, the recognition of "qualia" (qualitative facts that cannot be reduced to quantitative relations), and synergisms. Many synergisms turn out to be special laws of nature, not just special applications of the general natural laws.

After these general themes of intellectual limits, we will proceed toward some positive debate of what can be known. First, there will be a vision of the world as consisting foremost of designs, while matter and energy reduce to a qualityless dimension, analogous to space and time. In consequence of this we will proceed to develop the concept that change in the world can be viewed as a set of processes: designs in the time dimension. Thus, the general concept of causality can be made to subsume the energy concept into an understanding of the process element of existence: designs in a dynamic framework. Explanations of cause and effect are shown to consist essentially in descriptions of invariances. Sometimes these can be reduced to special cases of larger, more general invariances, but sooner or later, the analysis has to stop short at a stage where explanation equals description.

Building on such reflections, an attempt will be made to show how reality can be understood as mind—a revival of the concept of *logos* that was given to us by Greek thinkers of antiquity and that has been developed many times since then. The monism of materialism can be inverted: It turns out to have been due to a false perspective.

These several ideas are thereafter applied to some large yet specific problem areas—the controversy about evolution versus creation and the problem of scientific inquiry into human affairs. Both of these lines of discussion should help in identifying pseudoscience, as well as those areas of life and experience where science, properly understood, is not applicable. There is more under the sun than can be explained by anyone's philosophy.

Pseudoscience merely reflects someone's state of mind. It is the most widespread in the studies of human affairs because these are the most complex of all known reality.

We will raise many more questions throughout these essays than we will provide answers. Most of the conclusions we offer as alternatives to current science are framed as hypotheses. The starting point of it all has been the observation that ignorance is underestimated as a factor in our thinking and our cultural life. Acknowledging our ignorance is often as important as any claim to scientific knowledge.

An attempt will be made in the final essay to sketch a realistic program for science in society as it exists. It will be not as any kind of "final solutions"—that would be contrary to the main thrust of this book—but as illustrations of the ways in which we may have to meet the perpetual dilemmas of knowledge, belief, and faith.

TWO

Lighted Room: The Ways We See Things

Oculis acutis nox pervia
(To sharp eyes the night is accessible)
Motto of Accademia dei lincei, Rome, 1609–1632

You are in a lighted room at night. The light provides easy use of the room and its contents. They are all adequately illuminated or at least they can be. But the light fills only that one room—or maybe other rooms in the same building, many rooms if you will. The outside, by contrast, is poorly lighted, if at all.

For simplicity, assume the outdoors to be in the country or in a residential area without any street lights. Looking out the windows, you can see very little if indeed anything. If the lights in the room are very bright, all you can see in the windows may be reflections of the room itself. The ultimate in indoor lighting renders you "outdoors blind," unless, of course, you happen to be night blind to begin with.

To see some of what goes on outdoors, dim your indoor lights. Better still, put them out altogether. Adjusting your eyes to the faint light we call "dark," you will discover many things to be visible out there. The world out-of-doors is much larger than your room or even your whole building. Even with the sparse light of a moonless and lampless night, there is, in total, much more to be seen outdoors than indoors. It all depends on purpose and on training. The invention of night goggles, used by the military for instance, shows how the clarity of vision may be relativized.

BRIGHTNESS CONTRAST

The phenomenon is well-known to experimental psychology. The German term is *Helligkeitskontrast*. It goes from the trivial to the grandiose. On the trivial level, you may recall a commercial where a housewife says she thought her husband's shirt was white until she compared it to that of so-and-so, which had been washed much whiter in the detergent being advertised. On the grandiose level, there are, for instance, the dark sunspots that look dark against the surrounding solar surface. In fact, they shine with a brightness that is thousands of times brighter than the full moon; however, they are out-shone by the even brighter solar surface around them. The full moon, of course, looks very bright in the deep of night. It becomes impressive because in a moonlit night, the shadows all appear deep black and there is no penumbra as in a sunlit landscape. In daytime, the moon may be barely discernible as a pale white speck in the sky, no more striking than a minor cloud.

The lighted room is a powerful simile if we want to understand many of the workings of the human mind, especially its limitations. The things we can see and do often define what we cannot see or do. This simile can be used to explain the narrowmindedness of specialization, all the way from parochialism, on to the *acedia* of the one-sided scholar. The same simile can serve to explain indirect perceptions, choice of focus because of bias and purpose, the relativity of time concepts, of emotions also outside the sphere of direct human experience, and of consciousness, that ultimate enigma of our existence. The same simile can help us cope with some of the difficulties of knowledge beyond common sense.

SELF-LIMITING VISION

The window, which only reflects the room we are in, has many parallels in everyday living. Some psychologists will tell us that we can only see what we have seen before. An entirely new phenomenon is not really "seen" all at once, not until we have focused on it repeatedly so that its unusual features begin to be part of what we remember and are able to reconstruct in the receiving centers of the brain. Thus our vision is heavily influenced by what we have seen before. This should help dispel the often repeated fallacy that "a picture says more than a thousand words." If the picture says anything, it is because we already have words to articulate it. There are some good examples from war propaganda to show that, when placed in a different context, an old picture can be made to support a story that is entirely different from its original context. A picture of a giraffe would have told the old lady even less than did the living specimen in a zoo, because "there ain't no such animals."

The basic difficulty of perceiving what is not familiar underpins the cultural trait we call parochialism. This extends all the way from "the forest of weeds" (Hans Christian Andersen) as the only world the snails need acknowledge, to narrow nationalism and regionalism. The story of Noah's ark (or Ut-Napishtim's boat) told of a flood covering "the whole world," when in fact only the Euphrates-Tigris basin was flooded, as archeology attests that it was, indeed. However, this river plain, immensely vast to its early inhabitants, was all the world the chronicler cared to know about—it was the whole world to them. Similarly, one of the early rulers in this part of the Middle East declared himself to be "king of the four directions of the sky," because he was master of the whole twin-river region—again, the whole world by his perception. What was outside was as obscure as any unlighted outdoors. Only rumors may have told of the remote areas' barbaric and irrelevant inhabitants, as shadowy to Mesopotamia as "East of Eden" to the Book of Genesis. The parallels are many throughout world history, as, for instance, use of the French expression *tout le monde* (all the world) when only high society was being referred to.

The extreme of self-limited vision is in the scientific specialist at work through microscope or telescope. A typical case is a laboratory biologist working with an electronic microscope who never noticed that the squirrels around his suburban home moved by jumping. Focusing on either the magnification of the extremely small or the encompassing of the unfathomably large, instrumental specialists easily become victims of their own prowess. The sharper the specialty is defined, the more complete the focus on a limited area, the more there is that is left out—disregarded because of the general limitations of individual capabilities. It takes a large mind not to be lost in single-minded pursuit under sharp light.

There are analogies also among the trivia of modern day living, such as loud music and drugs, both providing a kind of overlighting that tends to outshine many real and worthwhile perceptions. Another example is the driver who does not understand that the glare of his windshield may make him unrecognizable to a friend he clearly sees on the sidewalk. Similar are cases of elderly spouses who no longer notice that their life's companions may speak with a different accent or dialect, or the professional psychiatrist who did not notice that his own young son was insane.

Recently, astronomers got an object lesson with the advent of radio astronomy. They now can see things outside the narrow band of wavelengths we conventionally call light. How many more heavens will there be to discover, in the sky or among living things around us, when we get other substitute "eyes" to widen our vision? Some animals have always been able to see some part of the infrared band, while others can see some bit into the ultraviolet.

Let us contemplate some cases where self-imposed limitations have damaged the vision of people in power.

IDIOCY OF CIVILIZATION

As a starter, we may dwell upon the phrase "idiocy of rural life" (*Idiotismus des Landlebens*), which we owe to Friedrich Engels, the much cited understudy of Karl Marx. In his condemning of the traditional peasant farmers, Engels vented a prejudice that was very specific to his time and to his mind bend. The same statement played a role in the development of Leninism. Lenin would acknowledge peasant farmers as people only when they had shed their rural prejudice and had begun to reason as Lenin wanted socialists to reason (F. Dovring 1996). The whole exercise is about as laudable as Richard Wagner's diatribe about the Jews and their music: The Jews would be all right if they would abandon their cultural heritage and become like good Germans. No real racism here, just cultural prejudice so thick you could spoon it with a fork.

The Engels quote reflects his ignorance (and that of many others, including Lenin) and reflects a lack of understanding of what was really going on in traditional rural life. We now know that traditional peasant farmers, living in a subsistence or semisubsistence economy, were among the shrewdest economic men ever. Detailed economic analysis has proven this to be so, but the main conclusion could have been reached by common-sense logic. Traditional peasant farmers were up against very harsh conditions, from both nature and society. It follows that any existing population of such peasants must consist of those who had stood the test of time: They had survived where you and I and Friedrich Engels would have perished.

It is time to realize that the achievements of civilization have been obtained with great sacrifices. The process is reflected in the early training of the human brain. It is now known that many synapses are abandoned during early childhood, rendering the growing and maturing individual more disciplined and one-sided than might have been the case under some less constraining system of education.

REASONS NOT TO SEE

The simile of the lighted room applies very much to intellectual development and history. During the Age of Reason in the 1700s, people rejoiced so much in the new-found light of science that they forgot about the kinds of wisdom still dwelling in the dimmer lights of popular culture. Thus was born the intellectual tradition that culminated in the self-blinded arrogance of early materialism: What cannot be proven does not exist. Pure thought turned into pure thoughtlessness. The ultimate consequence for modern dictatorships' use of disinformation was aptly identified by Orwell in *1984*: In the Ingsoc-Newspeak state, things not communicated did not exist.

It is by no mere coincidence that the Age of Reason followed on the heels of, and put an end to, one of the darkest episodes of Western cultural history: the flare-up of witch persecutions in the late 1600s. The new light taught that witchcraft did not exist. Thus a faith, a culture, a way of life inherent in lingering pagan practices was first obscured, then smothered in the rising light of a new dominant culture, that of conscious rationalism. Witchcraft was denied. The alleged witches did, of course, exist as flesh-and-blood human beings, but how they differed from approved, enlightened people was downplayed.

This is also how competent peasant farmers were deemed inferior to the self-appointed agricultural "improvers" of the age. They were just less advanced versions of the same basic people as had been perfected in the scientifically minded classes. We saw a parallel in the mid-1900s: Americans, the victors in World War II, tended to look upon the world's people as less developed because they had not yet reached the heights of American technology.

Protagonists of the Age of Reason sacrificed immense treasures of wisdom existing in the faint light of traditional culture. They did so in order to render secure the brighter illumination of the new culture's findings. The gains were proudly noted, while the losses caused by self-imposed limitations were ignored.

It has taken great efforts by cultural anthropologists and others—agricultural economists among them—to rediscover some of the wisdom that immature science had written off as superstition or as routines based on sheer ignorance. Peasant cultures now stand vindicated. With the means at hand in each age and place, they seldom had much leeway for doing otherwise than they actually did (and as some still do). In the sphere of material culture, the primitives know some things that advanced experts do not. This can stand as a prototype.

Training to become civilized includes some limitations, such as abandoning some primitive abilities, often sacrificing both originality and insight into nature. In the extreme cases, Communist ideologues in the Soviet Union tried to impose a prevalence of "consciousness" at the expense of the more primitive "spontaneity," which might escape rational analysis. The superiority of free enterprise over total social control reflects, above all, the creativity of the human mind, less available to rational analysis as that may be.

The fact that primitive people knew more botany than do most modern urbanites is interesting and is proven because the languages of the primitives usually included many more vernacular names of different plants than are found in the lexicon of most modern educated people. In the extreme case, modern education has failed to inspire students to learn outside of their areas of interest. When I asked a reasonably bright high school student if he knew the name of a tree on the parkway (e.g., a trivial species in the central Midwest), my young companion answered my question with a monosyllabic "no," but also showed no interest in the answer.

It is no great or novel discovery when schooling is found to mean suppression of many facets of an active mental life. At about the time when Friedrich Engels formed his ignorance of peasant life into a self-appointed norm, someone else in his country formulated the statement that "reading makes stupid" (*Lesen macht dumm*). School generally insists on forcing us to learn what we do the least well, not to excel in the things we do best.

SANITY AND GENIUS

The connection between sanity and genius has been belabored ad nauseam. Of course, not all geniuses are insane, nor vice versa. When many geniuses are more or less off the beaten track of the normal human mind, we must ask whether the problem may be in our concept of what is "normal."

Among many telling examples, let us mention two. Aldous Huxley and Eric Hoffer, modern thinkers of immense leverage, both had this (and only this) in common—that they were functionally blind during a stretch of years in their early youth. This made them escape the "disciplining" effects of conventional schooling and forced them to seek their lights on their own. And what lights! It is hard to imagine the contemporary world of thought without either "the doors of perception" or "the true believer." Neither could have been discovered under the normal light of the conventional schoolroom.

There are examples galore. Leonardo da Vinci was a motherless homosexual, as completely doomed to mental loneliness as anyone. William Blake never went to school and remained forever "unschooled," with precious gems in our cultural heritage rendered possible by this supposed deficiency. Alan Turing, the computer genius who gave his country immense wartime service as a code breaker, was later destroyed by that same country in the name of moral conformity.

In short, the abnormal genius is the norm of geniality (Berry 1992). The genius who appears normal is instead the exception that needs special explanation. We may venture the hypothesis that the average person is a potential genius, stunted by the effects of normal, or normative, education. One analogy is among the social insects, where the worker ants and bees are hunger dwarfs, rendered such by the educational and feeding regimes of the colony or the hive.

People who are mentally defective may, for that reason, escape the normal impact of conventional schooling, and so have the chance that their genius may be rescued from the mental "hunger dwarfing" of normalcy (Gardner 1991).

Why mental hunger dwarfing is the rule with average people is a lengthy subject that cannot be pursued here (F. Dovring 1984, Ch. 7). The analogy with the social insects breaks down because, in their case, the combination of many small bodies renders them functional parts of a whole of higher order, somewhat like the neurons in a brain—a "super organism" (Wilson 1992). Human individuals, already endowed with enormous brain capacities, are rendered poorer by the combination into society, which fails to create any reality of higher order. Very briefly expressed, human societies introduced this stunting social discipline as an instrument of warfare, rendered necessary (or so it seemed) by the chronic pressure of overpopulation, which the "ecological dominant" Homo sapiens did not see relieved by those automatic means that keep the rest of the Gaia system in balance. How different societies throughout time have coped with this overarching problem of human existence is a subject to fill a whole book.

RATIONING THE MIND

The spectacle of widespread stunting of human minds has called forth a reflection recently made that the brain, far from being the origin of consciousness and mental life, rather serves to ration them, by internal censorship (Darling 1995). The abandoning of many brain synapses during childhood then means abandoning many alternative paths of mental life, a pruning away of countless opportunities for individual development and creativity. The human anthill does not need many geniuses; those we already get *quand même* are bothersome enough. The troublesome effects of overheated technology development give seeming justification: Hurry but slowly, please, else the whole fabric of our society, and of the home planet's ecology, may blow up. There is no easy answer. Obviously all the stunted minds cannot be freed all at once, not until new ways of living have been discovered.

In one sense, we must now prepare a change of direction. Civilization meant not only inability to comprehend what goes on in minds otherwise schooled than our own. It also meant sacrificing much of the intuitive wisdom found in many primitive cultures. High on the list of what should be rescued from the destructive zeal of civilization are the remaining Australian aboriginals, who evidently knew/know things of the human mind that most of us, like Engels, have lost sight of (Morgan 1991). The totality of primitive experience includes more than the gene pool of primitive crops, which the plant breeders now realize are indispensable for their work in the future. All the human insights of, say, Western Hemisphere "Indians" and of tribal Africans down to the remaining bushmen in the bush must be carefully observed and protected to preserve and give new life to mental accomplishments that have been plowed under for most of us by school, military drill, and economic discipline ranging from office routines to the routines of speculative business.

VERNACULAR

An interesting case can be made of civilized language versus popular speech. There is a telling passage in Gogol's *Dead Souls* (end of chapter 5) where the author compares peasant Russian with the literary languages of England, France, and Germany. None of the latter comes anywhere near the earthy expressiveness of Russian, says Gogol. This is on account of a Russian peasant character's boundless exuberance in the use of derogatory idioms. Here Gogol compares what should not be compared. The English, the French, and the German languages as Gogol knew them were those of literati, civilized and shorn of the primitive expressiveness that no doubt still existed in the West European boondocks to which a traveling Russian literate had no access. Literates such as Engels and Lenin just condemned those boondocks for their "idiocy." It is telling that, in praising the Russian language, Gogol also in passing disparaged the language of the *vepsy*, the Finns of the St. Petersburg area, who still spoke an idiom close to that of modern Finland. We have competent testimony of Finnish to the effect that it is a very expressive language, extremely capable of shooting off those derogatory salvos against fellows on the road that Gogol found so admirable in Russian. As usual, the unknown is underestimated.

Under an even wider horizon, we may question whether the much-praised accomplishment of Homo sapiens to be able to *speak*, in ways no other animal can, may likewise have been obtained against the sacrifice of animal capabilities in communication and emotional expression. That, unfortunately, may come in the category of things we are systemically incapable of ever knowing about.

BLIND AS A BAT

Let us contemplate an extreme case. Bats are blind in that they cannot see as we do. Does that make them "inferior"? When it comes to catching mosquitoes in the dark, they are clearly superior to each and every one among us conceited humans. Not only that, but by catching insects in the dark, the bats earn precisely the kind of living they are designed for.

Bats are equipped with a kind of sonar like that now used in submarine warfare. By this gift of nature, the bats navigate skillfully in the dark of night and locate flying insects by the bouncing of their own sound signals.

Saying that bats are blind is no more to the point than it would be to say that those of us who do not have access to radio astronomy are blind because we cannot see those bands of wavelength (in continuous radiation) that are outside the range of conventional light. Those radio astronomers have a capability that most of us do not have. By analogy, we must propose that many traits of the vaunted superiority of Homo sapiens—other than the power to destroy—come from the limitations of our lighted room and our consequent inability to perceive things that many other sentient beings are able to perceive. Into this vast area of our ignorance we can throw some indirect light by contemplating the entire scope of subjective knowledge, which we will do in the next essay. As a lead-in, let us briefly sketch the subject of consciousness.

CONSCIOUSNESS

This is often represented as being a huge problem (e.g., Dennett 1991), when in fact it is trivial experience. The existence of consciousness is proven—if proof were needed—by the fact that we know the difference of being unconscious or in a dreamy state of subconsciousness.

The trivial fact of our human consciousness is best understood as a matter of clarity. The neocortex in our brains outshines any lower-order consciousness that may reside in less elaborate organs of the body. The analogy is with the sun outshining its own "dark" sunspots, to say nothing of the moon.

Is the brain a colony of bacteria? The case has been made not long ago, in a discussion of advanced biology (Margulis and Sagan 1986). If so, maybe the free-living bacteria in the primeval oceans were also conscious and communicating beings, on a level that our brains outshine hopelessly. Then maybe the primeval oceans, for a billion years or more, were the home of a vaguely focused brain. It may still have its analogue in the oceans of today, with their much more richly varied, but still intercommunicating, aquatic life. The fish were supposed to be mute because they had nothing to say. In fact, the fish have been found to communicate intensively across the expanses of their watery homeland (Earle 1995).

The problems of consciousness—if problems they are—will be discussed further when we come to the subjectivity of pain and other emotions.

THREE

Just Inside: The Ways We Know Things

Cogito, ergo sum
(I think, thus I exist)

René Descartes, early 1600s

In the modern world, knowledge is usually equated with the cognizance of objective truth, objectively established or objectively proven, or apparently so. Specifically, scientific knowledge is recognized as such only if it can be tested by accepted scientific techniques—techniques capable of distinguishing what we can prove to be so from what we cannot (for now, at least) prove to be so. Recent references to paradigms as guiding lights to discovering new sets of scientific truths appear to relativize the concept of scientific knowledge: In many cases, scientific findings are found to hold only to the extent the assumptions of an underlying paradigm can be upheld.

Everyday experience tells a different story. There are many ways in which we "know" things without conforming to the strictures of science. On the borderline, there are the *axioms*, statements that cannot be proven but are accepted because they are self-evident and indispensable. All science ultimately derives from self-evident, axiomatic statements. Without them we could not build even an elementary edifice of knowledge. The axioms are like the ground we stand on. Give me a fixed point outside the Earth, and I can move the very planet. But standing on the Earth I cannot move it, nor do I have any reason for wanting to do so. Axioms are accepted as self-evident so long as nothing (nothing well-known, that is) contradicts them. Should a conflict arise, the axioms may have to be reformulated, but we still need them as foundations.

Other borderline categories are where our knowledge is predicated on the nature of our instruments of observation and how these may relate to the needs and objectives of the people (or whomever) that depend upon these instruments for their perceptions. The previous essay dealt with what we may call "the lighted room syndrome." For now, let us contemplate the many cases of statements that are unproven yet accepted. Following that, we want to get into the vast areas where insights have their meanings only in relation to some subject. Intuitive image recognition, subjective time, consciousness, and emotions are important examples. Venturing further into animal feelings, we also want to explore the questions of "truth" in beauty and faith and of "truth" on ego versus mental field.

UNPROVEN YET ACCEPTED

The large, traditional problem is that of common-sense knowledge. From the Socratic teachings we have the distinction between *doxa* and *episteme*. Doxa is defined as an opinion that may very well be correct and often is, but cannot be proven so, or at least has not yet been proven so. Episteme, by contrast, is knowledge that we know to be true: We have the means to prove it and we have used those means. That cannot be said about the doxa type of statements, although they turn out to be numerous in daily life and of great practical consequence.

There is more to the doxa concept than mere lack of evidence for the time being. On the one hand, many doxa statements could be brought to the scientific test—can the statement be falsified or not?—if we had time to do so. Others cannot be so tested, yet cannot be rejected as irrelevant either in our daily practical affairs or even in our thinking of what really goes on in the world.

Among the facts of daily life is the fact that we are surrounded by a profusion of matters on which we hold opinions without being able to prove them, if for no other reason than lack of time to do so. If we were to subject every detail of what goes on around us to proof by scientific criteria, then we would not have time to live our lives. Recent evidence about the fuzziness of many things only renders this problem even more ingrained (Zadeh and Kaprczyk 1992; Kosko 1993; McNeil and Freiberger 1993). Even if we had all the evidence, to act upon all of it we would have to find the time needed to secure and to test this mass of evidence.

Most of the time we could not do that, as a practical matter. Short of scientific evidence, we often must act on what we "feel" or believe because no action at all may be worse than whatever we decide to do based on our unproven knowledge. In many cases, "no action" is, of course, also action, if implicit, and it may determine what follows as certainly as any action does.

In modern times, many people try honestly to raise the quality of the data on which they must rely by criteria they believe to be scientific. This is not always an improvement. Most people have a weak enough concept of science to make them lend more credence to their own supposedly science-based conclusions than these conclusions deserve. Going on untested common sense, "horse sense," may in fact often be less hazardous than substituting some findings from sloppy science.

Some of this concerns widely held misunderstandings of "probability," as well as scientific approximations. Many historians, for example, have been guilty of the logical fallacy that something that is, in general, very probable should also be accepted as true in a case at hand. They overlook the fact that probability is a statistical concept, not a final statement about any particular case. Saying that such-and-such has a 99 percent probability of occurring does not say anything about an individual case at hand; nothing prevents that case from being the one hundredth case. Sometimes this has to be so. The chance that it be so is, in fact, indeterminate.

Scientific approximations are not always well understood either. Even though we know that the Earth rotates around the Sun, there is no harm in referring to when the Sun rises and sets. It only appears that way, but using the inherited, common-sense reference to sunrise and sunset does no harm when we know the real nature of what goes on. Saying instead "when the Earth on our longitude turns toward (or away from) the Sun" may be more accurate, but it is also more obscure. Similarly, when Einstein showed that Newton's formulation of the law of gravity is an approximation that is not adequate in certain astronomic applications, many modern people jump to the conclusion that Newton's law has been disproven also for the fall of an apple. This is in fact not so. Newton's formulation is still quite satisfactory as an approximation of what goes on here on Earth, where the distances and the quantities are small enough that Einstein's formulations make no difference that we have any reason to be interested in. Einstein's formulation itself presents some minor unexplained problems on the cosmic scale, in some very exotic circumstances. It has still not been shown how a force such as gravity can act across enormous distances of astronomic void, as well as over short distances all around us.

Beyond the needs of daily life to use approximations or common sense derived from experience, there are other forms of subjective knowledge that we must recognize, lest much of what we debate in the name of science become self-defeating by way of logical circles. We may cite the cases of image recognition, inner time, and measurement of emotions— ultimately the use of different ranges of sensory and other inputs to practical knowledge. Astride it all is the problematic of consciousness.

IMAGE RECOGNITION

One case of indispensable doxa is that we recognize the face or the posture of a person of our acquaintance. "I could tell that face from a million." Yes, we recognize that face, but can we prove the identity, in any strict scientific sense? Can we objectively specify all the criteria for telling that person apart from some close look-alike? We might be fooled, especially if it has been a long time since we last saw the person. If it was relatively recently, our recognition of features, posture, and so forth may be strengthened by something as subtle and hard-to-define as mental expression, which may also include the signals of mutual recognition in eye contact. If the other person meets our recognizing greeting with a blank stare and a refusal to reciprocate our greeting, this may call into question the identity we thought we had established. At the same time, even the most positive recognition may be complicated by conflicting motives. Recognition in a line-up of criminal suspects is the extreme case of where we may have to admit some uncertainty—there have been spectacular cases of traps deliberately set. Recent advances in computer application to image recognition may often help, but they seldom, if ever, are enough to eliminate all doubt, based as they are on approximation and probability.

There are, of course, borderline cases where we have a use for scientific criteria to validate or invalidate a subjective identification of some uncertainty. Some criteria can eliminate a suspect, for instance by fingerprints, blood types, and genetic patterns. Eventually, science may be able to identify the brain mechanisms that allow us to recognize a face (or some other complex phenomenon) just at a glance. This may or may not come to pass. Any such scientific procedure will, in any event, be much too cumbersome for it to substitute for common-sense recognition in daily life.

The point to make here is that, in the great majority of day-to-day events, we would not have the time to await scientific validation of our pattern recognition, even disregarding any continuing imperfection in future scientific findings and the techniques based on them. Immediate recognition is nearly always followed by renewed relations between people, whatever their relations may be. Doxa simply has to be sufficient for trivial practical reasons. The same goes for many other types of pattern recognition, such as that of plant or animal species or varieties, brands of automobiles, or styles of architecture.

INNER TIME

The concept of inner time was proposed by Alexis Carrel (1935). He stated that, to an organism, the relevant time concept is measured by what goes on in the organism.

Usually, we think of time as absolute, objective, and real. Its units are derived from the movements of our planet, the Earth. One revolution around its axis becomes a day-and-night cycle (*sutki* in Russian, *dygn* in Swedish). One revolution around the Sun becomes a year, a unit recognized in all languages. The complications stemming from the Sun's own movements, which cause the Earth's movement to be really spiral, seldom bothers anyone, not even the astronomers who count distances in light-years. Much less is it explained how this spiral movement is further complicated by the movement of the galaxy (the "Milky Way")— or, rather, a cluster of galaxies to which it belongs—within which the Sun moves in a kind of elliptical pattern. The Milky Way has recently been found to move at great speed in a direction that is assumed to be toward some otherwise unknown "Great Attractor" (Dressler 1995). That movement of the galaxy thus becomes an elliptic spiral, within which the Earth makes its own spiral. None of these relativities seem to prevent us from using the year as an absolute unit. That should tell us something about the relativity of inner time.

Inner time, says Carrel, depends on what goes on in the organism. In childhood and youth, our years appear longer because more happens in them. Old age, by contrast, has years that are shorter and shorter. Our time has come to an end when we grow so old that nothing happens inside ourselves anymore.

The concept of inner time has come to renewed life by recent discoveries of internal clocks in mammals. According to these findings, an elephant (for instance) that lives to the normal life expectancy of elephants experience about 800 million heartbeats and 200 million draws of breath in a lifetime. So does a mouse that lives to the normal life expectancy of mice. That life expectancy is much shorter than the elephant's, of course, so that must mean that the heart and breathing of the mouse is correspondingly faster. The same, we are told, holds true for nearly all other species of mammals. The most important exception to this rule is our species, Homo sapiens. In our normal life span of seventy-five years (or so), we experience about three times as many heartbeats and draws of breath as elephants or mice do in theirs (Seielstad 1989, Ch. 3). This really points to the human species as a serious innovation, away from the ordinary pattern of mammals. The deviation may be even larger, for there is a question about whether our seventy-five years are the real measure of what a human life span could or should amount to. Some out-of-the-way mountain tribes (the Hunza of Cashmere, the Abkhazians of the Caucasus, and at least one tribe in the high Andes of Ecuador) appear to have much longer life expectancies, and so do the aboriginals of Australia. Maybe most of us are stunted by some undiscovered incidental (Aldous Huxley mentioned an ingenious hypothesis, in *After Many a Summer . . .*).

Let us return to the mouse and the elephant. Having about the same number of heartbeats and draws of breath in a normal life span should really mean that these animals, in some subjective sense, have the same life span. Small animals generally live shorter life spans than large ones, but they may indeed be living faster. Try to follow with your eyes the movements of a squirrel out in the garden or a park and you get an impression of someone in a big hurry. Time is short! But that is our anthropocentric mistake. To his own senses, the squirrel does not hurry any more than we do on our time scale. Having more rapid rhythms, small animals may, in some sense, accomplish more in a day or a year than do large animals. The suggestion was made recently that many birds, for instance, may be more intelligent than is suggested by the small size of their brains, because in a short life span they may use these brains more intensively than we use our (Barber 1993).

It is a different matter that the human brain, that impressive array of some 100 billion neurons, may be underemployed—it may represent an extreme case of what economists call "excess capacity." The study of hydranencephalic individuals, who may have only one-tenth as many gray cells as most of us, showed that they still managed to develop intellectually as well as anyone (Darling 1995, pp. 82–84). This definitely points to large excess capacity of brain cells in the normal human being.

That argument could be developed a great deal more. For now, let us only underline, once again, that perception of inner time is an important type of subjective knowledge, essentially out of reach for scientific analysis.

This may be necessary as a counterpoise to the recent rather vocal debate among scientists as to the "nature" of time: cyclical or "arrow" (Coveney and Highfield 1990; Gould 1987). The arrow concept is particularly cultivated by those who debate the "Big Bang" theory and the theory that the age of the Cosmos is finite. That, apparently, disregards the possibility that the cosmos being debated is *our* cosmos, that there may be others like it at unspeakable distances in both space and time. The cataclysmic conception of the Big Bang underlies the formulation of the "edge of time" (Gribbin 1992; Halpern 1992). On a scale of many other cosmoses the arrow of our cosmos might be no more than a seemingly straight, but relatively very short, segment of a much larger cycle.

Any such reformulation of the large time problem is, of course, of as little human interest as the cosmological time concept itself. On our scale of being, the cosmological problems may be as they are; they do not really concern us in the way we try to conduct our lives.

EMOTIONS

We cannot measure emotions, yet they are real. This is a vast area of subjective knowledge that in many ways eludes scientific treatment, for in the minds of most scientists, the epitome of science is measurement—quantification. The subjectivity of our emotions is such a stumbling block to scientific inquiry that some scientists—who should know better, really—even have taken it upon themselves to declare that animals do not have any emotions; they are mere mechanisms that we can handle any way we choose to, without being guilty of any act of cruelty.

Let us begin with pain, the simplest and most straightforward of the emotions. We all know that it exists and has (or can have) powerful sway over our lives. We all know that it can be weak, moderate, strong, or beyond endurance. Yet we cannot measure it. Even weak pain can be hard to endure if it is unending, relentlessly throbbing, hour after hour. Some people get used to it, others never do. Manufacturers of pain relief medications compete in stating how strong their brand is, implying how it can relieve strong pain.

Asking around in the medical professions, the answers are unvarying: Yes, pain is real, and no, we cannot measure pain.

That pain is real is intuitively clear to nearly everyone. However, there are some self-professed exceptions: The Stoics and recent believers in Christian Science maintain that pain is an illusion. But since we are agreed that pain is subjective, there seems to be little to quibble about. The reality of pain can, however, be confirmed objectively by the concept of "referred pains." These are pains that originate in one part of the body, but are experienced, subjectively, in some other body part. For instance, something may be wrong in the colon, but the pain is felt in the lower part of one of the legs. The location patterns of referred pains are an important diagnostic tool for medicine, especially that of the interior organs. When referred pains serve diagnosis, they must, of course, be real. They are not just something the patients say that they have; they cannot be mere illusions. Their distinct location patterns and how these relate to distinct medical conditions that are found objectively to be real testify to the reality of the pain itself. The reason for such "referral" of pain seems to be in the way in which the pain signals travel from the afflicted organ by way of peripheral nerve connections before they reach the brain, the putative center of consciousness.

The same also confirms the objective existence of consciousness, if that were to need to be confirmed. An unconscious person could not report on referred pains, or on any pain, except maybe under hypnosis.

But why cannot pain be measured? Why indeed? To measure anything, we need a yardstick. Each of us has one, of sorts; we can describe our pain as weak, moderate, strong, or beyond endurance. But we have no assurance that this yardstick is the same for everybody—indeed for anyone other than ourselves. It may not even remain the same for an individual over time. Some pains are weakened by repetition; others may become worse, even unendurable, for the same reason. Anyone saying "my pain is worse than yours" is talking nonsense. The same objective cause of pain—say, a tooth abscess—may cause pain of different magnitude to individuals who are differently susceptible. Among other things, the ability to focus attention on something else can render pain more endurable, that is, subjectively weaker. Egocentric individuals, for instance, may focus their attention precisely on their pain, hence rendering it worse, and vice versa with individuals who are able to focus attention on something other than their everlasting aches. Subjective experience of pain may thus reflect character as much as it reflects objective medical conditions. The pains are no less real for that sake.

The same is true of pains and painful experiences that do not reflect physiology, but some other of life's circumstances. Often repeated is the question about soldiers at a war front versus relatives back home: Who suffers the worst, really? No one can tell, again, because human suffering includes the consequences of character no less than those of objective misfortunes.

What goes for pain goes for all the other emotions—less obviously, maybe, but no less surely. Feelings of well-being (the opposite of pain) are more vague, but no less real for that sake. Joy and sorrow, elation and depression, relief and horror, a sense of beauty and of ugliness, love and revulsion—all are real for each individual. All are equally inaccessible to observation or measurement by scientific inquiry. Some of the consequences of our emotions may be possible to approach scientifically, but indirect or implicit inference is not the same as scientific knowledge. Among the worst fallacies that are bandied about nowadays is the statement that "If I can control myself, so can you." The strength of emotion cannot be compared from one individual to another, nor can the ability of self-control.

ANIMAL FEELINGS AND CONSCIOUSNESS

The subjectivity of feelings or emotions lets us resolve the enduring problem about feelings and consciousness in animals (Griffin 1984). Any pet owner knows, of course, that their pet can experience emotions, pain among them. But how far does that extend down the ladder toward "lower" forms of life? The only reasonable assumption is one based on contemplation of the lighted room syndrome. It must be assumed that each species is capable of experiencing pain and other emotions on its own scale and by its own lights. Take the extreme case of pain to the limit of endurance. Pain is a practical matter; its mission is to warn against dangers to be avoided and ills to be remedied. Without pain signals as warnings, animals would not be able to cope with many of the situations they must cope with for survival. How else would pain function, if it were not conscious? The question of whether or not animals feel as strongly as we do is, therefore, meaningless. We cannot measure our emotions or compare them in any strict sense. But we can say when pain, or any other emotion, gets to the limit of what we can take ("It is so beautiful that it hurts."). By analogy, we must assume that any animal can feel pain to the limit of its endurance. Such pain must be as severe to them as it is to us. Suffering to the limit of endurance can happen to the dog and the insect and the earthworm. Each organism measures its pain by its own instrumentation. Pain beyond endurance must be subjectively as real to an earthworm as it is to you or me.

The consequences for animal protection should be clear enough (Raymo 1991). We have no reason not to feel empathy with everything living. When we have to kill, as we often have to (e.g., in pest control), we should do so as swiftly and mercifully as we know how. Gratuitous infliction of pain will hurt our own humanity, unless we have already shut that kind of insight out of ourselves—in which case we may already have badly hurt our humanity.

Animals need more than the absence of severe pain. Joy and love are as real to them as to us all, according to our lights, our scale of perception. Happiness can be seen not only in ordinary house pets that are properly kept. The famous Lipizzaner horses of the Spanish riding school in Vienna form an effective contrast against the majority of horses that have been trained by the whip. The Lipizzaners are educated by pleasurable inducement. Recently we have seen many engaging scenes of people playing with dolphins, the huge water mammals evidently enjoying this activity. There was a powerful message in the recent film on Leni Riefenstahl: The aging actress, in scuba gear, was seen patting a sting ray on the back, with the huge and dangerous sea beast accepting this in a relaxed posture, evidently enjoying this contact with a living being from beyond the world of water.

ANTHROPOCENTRISM

Most of the literature about animals versus humans dwells on the traits that make humans in some sense superior to animals. In practical terms the superiority is obvious enough, but precarious: The "ecological dominant" is able to sweep away any obstacles to our immediate progress, but the long-term outlook is far from clear.

Much less written about is the loss of animal abilities that took place when Homo sapiens ceased to be just another animal species. The growth of the neocortex in the brain and of the resulting thinking capacity—far from used in full—went along with losses of acuity in the senses. Similarly, the sharpening of the higher (more specialized) senses, such as sight and hearing, in the higher animals also went along with a dulling of the more vague sense of the whole organism. On both these levels, we witness a sharper light in the specific room while receving less and less from the outside. We must question whether people are really more intelligent than animals and whether the higher ones are more so than the lower ones. The primeval ocean full of bacteria may have brimmed with sensory life of a type and an extensiveness we can only dimly imagine. Fish communication in the ocean waters may be another analogy to the brain worth pondering.

We may well ask whether there is any merit to the talk of nature's cruelty. To the contrary, it appears that many predators may have the ability to induce their victims into a stupor so that they die without feeling anything. If so, it may very well be that Homo sapiens is the most cruel being on Earth, followed closely by the chimpanzee, our closest cousin.

"IN THE EYE OF THE BEHOLDER"

This time-worn statement is generally taken to mean that beauty has no objective existence; it is only functional (or "virtual") in relation to those who appreciate it. The case of the tone-deaf person who thought music lovers stupid could be a touchstone.

Some beauty can be established in an objective way. Musical harmonies either are there or they are not. A jarring note is ugly by any standard. It can have its emotional, artistic (e.g., dramatic) reason for being retained in a work of art, but beauty it is not. The same may extend to an artistic conception taken as a whole.

The generally assumed subjectivity of beauty may be dissolved if we take the time to contemplate how different people are in their limitations. Congenitally, we may all be hereditary geniuses with the potential (at birth) to develop into some superhuman like the Panurge of Rabelais' *Gargantua* tale. As a practical matter, our upbringing renders us (or most of us, anyway) severely stunted, able to develop only certain limited portions of our original, nature-given endowment. Therefore, our ability to appreciate beauty may have been narrowed down into certain selective "bands," analogous to the bands of light our eyes are able to see. Unlike sight, those abilities to appreciate what is beautiful or artistically effective are not limited for all humans in the same way or to the same extent. We may have different "windows," depending on what brain synapses were disconnected early in life. Accordingly we may be able to perceive only what comes into this mental field of vision. Disagreements as to what is beautiful may therefore be subjectively honest yet objectively wrong. Not only are the tone deaf wrong in declaring the lovers of music to be stupid; placing oneself as sole (or principal) arbiter of beauty, and condescending to tell others about it, is therefore wrong on principle. Experts on art and literature endeavoring to judge other people's tastes are, essentially, a bunch of humbugs. Objective judgment would assume that these individual experts are bright enough to perceive all the beauty in the Cosmos. Hardly any mortal can be trusted with having all that vision. Our lighted rooms are much too confining for that.

A similar kind of relativity as governs our access to beauty may also hold the key to all the individual varieties of religious experience. Touching on this immensely sensitive area of discussion, we need not try to propose any opinion about the possible objective validity of religious experiences and convictions. In the context of what we are discussing, it may be sufficient to acknowledge that religious experience is widespread among people and often leads to religious faith, and that both the experience and the faith often have powerful consequences for the lives of individuals and societies alike.

Religious experience and religious faith, when they are genuinely experienced and genuinely felt, are known to be powerful agents of mental healing. Dismissing them as scientifically unproven—nay, unprovable—is, therefore, not all that can be said about them. What we would stress here is the great variety of such experiences and faiths. This variety seems, like the individual sense of beauty, to take, from a vast storehouse of opportunities, those that fit the needs and the capabilities of the individual. The major religions seem to confirm this by the latitude they offer individuals in their choice of specific experience and their specific attachments to symbols and articles of faith. This is so in all the world religions in modern times, and this was, of course, so also in the pagan religions of the Mediterranean region and Europe. What really may lie behind it all is certainly beyond the intent of this essay.

Something similar, individually different "windows" capturing individually different parts of a vast reality, could also be thought of as the key to the bewildering variety of some people's experience of transcendental or parapsychic phenomena. Filtered through the minds of some individuals, such phenomena seem to escape the methods of objective science. For all that, they could still be real in some sense.

EGO OR FIELD?

Among the obvious traits of subjective knowledge is our recognition of ourselves as existing. Descartes said it well, but the full meaning of our individual existence has been slow in coming.

Classical idealism viewed personality as something absolute and wanted to explain all our mental processes from this seemingly solid starting point. Some extreme thinkers were driven toward "solipsism," where one recognizes only one's own ego as existing while all else— other people, other things—are just subjective experiences of this absolute ego. We owe a classical formulation of this line of thought to one Max Stirner (ca. 1850), whose masterpiece was called *Der Einzige und sein Eigenthum* (*The One-And-Only and His Property*).

Such extreme "philosophical egoism" is not widely accepted, however. For most of those who tried to comprehend what was going on, the ego was relativized because of its ability to absorb information from outside itself and to emit information to other agents in the world. In other words, the ego, for whatever it might be worth in itself, is part of a wider field of designs and processes. Its limits are not sharp, rather they are "fuzzy" in the language of a recent school of thought coming from, of all places, electrical engineering (Kosko 1993).

The ego may be a "system" in the sense of modern systems theory, but an entirely closed system it is not. Each one of us is also a field or a part of one, or rather of multiple fields. Participation in adjacent fields is both strengthened and limited by the inner organization of the ego and its field complements. Assumptions of shared culture both help and hinder. Both aspects find expression in our use of elliptic communication, falling back for support on what is already present in the mental field.

Important are the shared impulses from surrounding culture. Unison singing in the churches has always served to consolidate the connections between individual egos. In our time, American popular music can be likened to brain waves that pervade the fabric of society and call forth analogous reactions in large numbers of people—not necessarily identical reactions, but reactions that are sufficiently analogous to be conducive to joint movements in new directions, as "in concert." In the extreme case, modern Western music was experienced in Russia of the 1960s as a "battering ram" that the West thrust into the cultural and social fabric of the Soviet Union, hastening its downfall (Gurevich 1991).

The nature of such connecting mental fields has been the object of various hypotheses, ranging from the "collective unconscious" (Jung 1938) to the "morphic resonance" proposed by Rupert Sheldrake (1991). Some of this may spill over into things we can never know, but that may be real just the same.

In a new extreme of interpretation, the ego has been branded as an illusion, to be dissolved only by the clarity of death (Darling 1995; Reanney 1991). Again, we are faced with the problem of illusion as reality. Virtual realities can be powerful enough. If we experience the ego, as Descartes did, then it must be real in some sense, if only functionally. *Only!* Whatever is functional can be very important in the world of realities. We shall have more to say about this when explaining what is meant by virtual reality. On the trivial level, for instance, money has no objective existence, it is a virtual reality that facilitates economic transactions. So let us not be too condescending toward "illusions." If the ego is an illusion, it is a mighty consequential one.

FOUR

Beyond Reach: Things We May Never Know

Fools rush in where angels fear to tread.

Alexander Pope (early 1700s)

Naive scientism tends to believe simply that all that is real can also be made known—be subject to objective scrutiny. The human mind, working correctly through scientific methods, could become master of all there is to be master of. Many of us no longer believe this. Let us see how and maybe why.

We already touched upon the theme of nature's intellectual traps as complicating the path of our progress. The standard case is that of radioactivity that had to be hard to detect on this planet because if it had been obvious, the planet would not have been suitable as a home for organic life. Hence we would not have been here to observe it. Knowledge of radioactivity became possible only with the advent of more advanced and refined instruments of knowledge.

Some obstacles appear to place knowledge more definitely beyond reach. In physics, the warning was issued some time ago: "Our intuition is based on our experience in the macroscopic world. There is no reason to expect our intuition to be valid for microscopic phenomena" (Hans Bethe, as quoted by Parker 1988, pp. 185ff.). Some recent astrophysicists have expanded on the same theme by proposing the concept of the "anthropic principle" of knowledge. The same applies, by analogy, to other domains of knowledge that present us with intractable difficulties.

Separated vastly from each other, physical cosmology and parapsychology both tend to rely on supports that, by definition, cannot be there. In both cases, something has to be the ultimate ground on which all experience rests and that refuses to be proven other than by circular reasoning. Like the ordinary axioms, such ultimate supports, if they are found, cannot also serve as their own supports.

Before trying to tackle these large themes of intellectual problems, let us look at some things closer to everyday life. Animal communication and the recently hypothesized category of "morphic resonance" (e.g., between animals) are examples of domains of possible reality where our very nature may deny us complete entrance.

ANIMAL TALK

Do animals talk? Mark Twain thought so in a delightful little sketch, "What stumped the blue jays." Blue jays are very vocal; they easily strike us as being talkative. People who are sensitive to the animal world often gain the impression that animals understand people better than people understand animals.

Can we talk to the animals? To some extent, yes (Griffin 1984). Anyone who has conversed with a cat or a dog knows that there can be some communication between people and other animals (Steiger and Steiger 1992). Eye contact appears to be important here. How far does this interspecies mental contact extend?

Not very far. By no means as far as that between people, incomplete as our understanding of each other may be. Recently there has been some very serious research into animal communication. For instance, chimpanzees and gorillas have been taught to some extent to express themselves by means of sign language used by humans who are unable to speak orally. Some degree of conversation between these animals and their trainers has taken place. It may be that anthropocentric illusions have made more of these results than they really warrant. Despite valiant efforts at avoiding anthropocentric traps, most of the research has centered around what animals can understand of us, not what they understand of each other (Candland 1993; Cavalieri and Singer 1993; Cheney and Seyfarth 1990; Goodall 1990; McCrone 1991; Morton and Page 1992; Peterson and Goodall 1993; Savage-Rumbaugh and Lewin 1994; Strum 1987). The problem is that most of the research asks what animals have of what we call language. There has been precious little attention to what animals really use to communicate with each other and what they may be using when trying to approach people.

In short, what is it that animals have and we lack (Griffin 1984)? There has been some progress in studying smell communication—a good example of where many animals are superior to humans. Dogs evidently get a great deal out of sniffing, but what it is that they get out of it we have not the remotest inkling about (Agosta 1992). Some things are now known about the dance-like acts by which honey bees inform each other about some of their lives' opportunities (von Frisch 1966). There is also an interesting amount of new data on birds, showing them sometimes to be surprisingly articulate in communication with people (Barber 1993; Skutch 1996). Small brain size may to some extent be compensated for by more intensive use of what brain there is. Short life span appears associated with a more rapid pace of functioning. But if there are means of communication in the animals' world that are essentially different from our own, we do not know what they are. It can be argued that we may be systemically incapable of learning about some specific forms of animal communication. It came as an eye opener to many of us when the proverbial "silent sea" turned out to be the emphatically "unsilent" sea, reverberating with signals between the fish of the depths (Morton and Page 1992). And this is no small thing: Most of the Earth's ecosphere is in its oceans (Earle 1995).

ALIEN SPEECH

To gauge what may be involved in animal talk, let us contemplate a couple of cases where human talk goes along unusual lines.

For instance, in the heyday of European colonial dominion of the world, a British administrator in a West African colony wrote to his government in London that the people (the Africans in the colony where he resided) were incredibly primitive. They had no real sense of time! Proof was that the language of these people lacked any past and future tenses. The colonial bureaucrat could not have been more wrong. The West African languages do have tenses for past and future, but expressed in a way the linguistically naive Englishman could not hear. Tense is expressed by musical tone. In this way, a shortcoming in the Englishman's insight became interpreted as a trait of inferiority among the Africans that he had not begun to understand. This was hardly excusable at the time. Europeans already had available plenty of knowledge about tonal languages—Chinese foremost, Thai to some extent. Even Swedish has some cases where tonality contributes to establish meaning.

This conflict was, of course, remedied in time, even if it took inordinately long before Europeans and Africans could begin talking to each other on somewhat equal terms. The feelings were often mutually derogatory, as in South Africa; not only did Europeans believe Africans less gifted for technical occupations (which is not true), but black Africans tended to regard whites as subhuman because they lack some of the emotional expressiveness of the blacks—again a case of underestimating what was not understood.

Another case. In Egypt at the time when defeating Israel was still on the political agenda along the Nile, there was a popular songstress calling herself Umm Khultum (same name as a daughter of the prophet Muhammed). She was famous for singing in a way only Arabs could appreciate. Western listeners thought she only sang one protracted note and found her boring. To ears attuned to the local culture, she produced minute vibrations that would send her Arab audience into emotional frenzy. To those not so attuned, this was a world as alien as the dogs' sense of smell.

Either of these two examples will do, but there can be any number of analogies, all the way to the tongue-clicking consonants of the Kalahari's Bushmen. As to tonality and small vibrations, it is not unreasonable to hypothesize that the doves' cooings, boringly repetitive as they may be to human ears, may be as expressive to other doves as Umm Khultum's singing to her Arab audience.

Other analogies may be, for instance, in the use of Morse code. The DNA in our genes are thought to work that way—there are only four chemical components, only four "primary symbols," if you like (Jones 1994), from which to build very complex messages in the genome. The Morse code used in telegraphy has only two primary symbols, the dash and the dot, from which to build analogues to the entire alphabet and more, by combinations of frequencies. Four basic chemical signs may lend themselves to many more combinations—more numerous by magnitudes. If we listen to the sparrows' chatterings or the crows' cawings, one cannot exclude the possibility that they may be full of very detailed expressions by the relative frequencies in which the various sounds are repeated.

To know any of these things, we should be able to break the codes in which the animals may be communicating. To do that is harder than identifying the components of dark matter in the cosmos. Breaking the code of human speech or texts requires a known language that has been translated into code. This was shown by the problem of the Minoan script from ancient Crete. The younger variant, Linear B, could be read because it is in a known language, namely Greek. But the older Linear A is in an unknown idiom and therefore remains without a reading, even though it obviously is in some human language. It must have meaning, but the meaning escapes us for lack of a key to the code. Similarly the Maya script from Central America and Mexico could at long last be read after many frustrating attempts, but only because some dialects still spoken in the area are late variants of the Maya language.

The above examples from animals' communication are much more difficult. Here we not only lack any key to the code, we also do not have access to the very minds that may find detailed expression in the animals' communication. Thus the most elementary precondition for code breaking is absent. Knowledge of the details of animals' communication is, as far as anyone can see, incompatible with the human mind. We are not just "superior" to the animals in many ways that we know of. We are also inferior to them in many ways about which we know much less.

Many animals are clearly superior to people in the art of parenting—on the whole they do it by suggestive communication, with little in the way of spanking. Civilization, which finds it commonplace to train horses by the whip, also has used plenty of senseless violence against the defenseless young. That can pass only behind the protective wall of the "ecological dominant," not among species that have to survive—and teach survival to their young—on the terms of total competition in the wilderness.

These remarks do not imply that we now know much more about animal communication than before. They only mean that we now can come to fathom the width of our ignorance on these matters, rather than insisting on "inanity" in animals (Pinker 1994). If they were so inane, how could they cope with life in the wild? Obviously, they are experts at survival, a very complex art. What the above reasoning accomplishes is to spell out the vastness of our ignorance about animal communication. Recognizing ignorance is as important as any positive knowledge.

MORPHIC RESONANCE

The term morphic resonance was proposed by Rupert Sheldrake (1991) to generalize a form of apparently nonmaterial communication between animals. He reported two types of observations.

One was of a rat (in New York, for instance) who found the shortest path through a maze by the usual process of trial and error; whereupon other rats, in other localities, found the same shortest way through the same maze without having to go through any trial and error.

The other case was of tits, in England and some other countries (the Netherlands was an important instance), who learned to steal cream out of bottles placed on people's doorsteps—the news of how to steal cream spread with amazing swiftness among the tit population. The case in the Netherlands was particularly intriguing because there the war interrupted this kind of cream delivery, and the postwar tits who learned the trick so rapidly were not even born before the war.

Sheldrake concluded that there is some form of communication from patterns not yet understood, but possible for some animals to pick up without having to go through conventionally known channels of communication or by aid of any known energy processes. Generalizing from such evidence, Sheldrake suggested that morphic resonance could be widespread in nature, eventually explaining many unexplained phenomena.

Many leaders in the scientific community have reacted to Sheldrake's general hypothesis with emotional rejection, bordering upon witch hunt. Sheldrake himself seems to have answered this treatment by withdrawing from scientific work in the conventional sense, concentrating his lights on problems of the environment. In all fairness, his results and his hypothesis ought to have been taken more seriously. It is in the nature of such things that the more they differ from the well-known paths of the human intellect, the more difficult they are to prove or to disprove. The parallels in animal communication should have commanded more respect for the unknown, some of which may remain unknowable forever. Whether this is so with morphic resonance only intensified search could bring to light. What if Sheldrake were to be proven right in the end? Recently he has come up with findings of dogs (and some other pets) knowing in advance when their masters will come home. This suggests some kind of mental telepathy.

PSYCHIC REALITIES

One basic difficulty was expressed calmly and lucidly by Mary Webb, a British novelist. Her "beat" was Shropshire, a region bordering Wales, where witchcraft and other occult forces have been a prominent part of local lore. In one of her novels there is a case of "poltergeister" (telekinesis), a psychic disturbance that has never been completely explained nor completely explained away. Yes, says the author's alter ego in the novel, such things do in fact exist, but we have no use for them—they just threaten to upset the balance of our mental powers, like a jarring note interfering with a piece of music.

The case is well chosen, for telekinesis is relatively well documented in recent serious investigations. The tendency of many scientists to reject out of hand all belief in psychic phenomena is analogous to some of the initial reactions to the first news about radioactivity. The explanations for such reactions are also analogous: If psychic disturbances had been rampant in our daily lives, then we would not have been able to function as rational beings in a society trying to be rational.

The basic difficulty is that psychic phenomena are supposed to be nonmaterial in nature. Then how could they be fitted into a worldview that assumes *some* material base for all real phenomena? For many alleged parapsychic phenomena—say, reincarnation—the problem is this:

Suppose there is such a class of realities with no demonstrable ties to "matter"—in other words, buildings without bricks—then the foremost problem is not even how such phenomena could exist. Rather, it is the problem: Suppose they do exist; how could we possibly have any direct evidence about them? All our means of evidence are tied to matter in one form or another. Hence our very apparatus for bringing into evidence is so steeped in space-time-quantum terms that it may be powerless to come to grips with realities that lack one or more of these terms—even quantum (Talbot 1986). The facile conclusion that psychic phenomena cannot exist is then based on circular logic. Even Occam's razor becomes circular logic in a situation like this.

All of this has not prevented some writers from stating that psychic phenomena are not supernatural—they have their own natural laws (Duncan and Roll 1995).

PROOF NEGATIVE

The naively scientoid standpoint has been restated recently (Stenger 1990). The categorical finding of "Psi = O" (alleged parapsychic phenomena cannot possibly exist) is so weakly founded that it resembles the Bible-based so-called "creation science," which has concentrated mainly on finding fault with selected features of accepted science. Stenger presents a thin anthology of arbitrarily chosen case histories. This does not by any means add up to any valid generalized proof. It is not enough to state that psychic phenomena cannot be derived from known scientific principles and facts or be established by any tested scientific techniques. For proof negative, one would first have to disprove all actual findings of alleged psychic phenomena, and that would take much more than writers like Stenger have attempted.

Further debunking has been tried from a psychology standpoint (Baker 1992). The emphasis here is on the complexity of the human brain and its estimated 200 trillion operations per second. The vast scope for synergisms in such a system appears overlooked. The writer is a dogmatic reductionist who even denies the reality of hypnotism. All unexpected mental states are brought down to hidden memories (the "cryptomnesia" of Flournoy). Occam's razor is used to argue that only the well-known can be true.

Characteristic is Robert Baker's approval of John Searle's comparison of the brain with the digestive tract (Baker 1992, Ch. 1). Polemizing against recent findings of "life review" in near-death experiences (more below), Baker stresses the discovery of a brain center apparently responsible for such experiences—not noticing the contradiction with evolution biology.

The reductionist credo is concentrated in the requirement that all credible phenomena must be capable of replication under carefully controlled experimental conditions. Such a formulation appears to deny the reality of any "flashes of genius" both in art and science. Such "bolts of clarity" have contributed immensely to the progress of science itself, but they cannot be explained, much less replicated at will, under ever so stringent laboratory controls.

Parapsychic or just psychic phenomena have, of course, been badly tainted by huge amounts of fraud that have been perpetrated by self-proclaimed mediums and their profit-minded promoters. But the negative experience of a million fraudulent cases could not erase the validity of a single case, or just a few cases, where paranormal experience cannot be explained away.

There are, in fact, more than just a few cases that defy negative proof. The continued resistance in many circles against all such experience, no matter how well established, therefore reflects something else and something more than sheer honest disbelief in face of the impossible. Nuclear reactions were also thought impossible before Becquerel, and the modern proof that the alchemists were pursuing a false lead (even Newton is said to have believed in alchemy) should not make one reject all chemistry. In contrast, an obvious case of false science such as astrology does not appear to attract as much negative polemic, despite the use of astrology by persons in high positions entrusted with important decisions affecting all of us.

There are interesting differences between classes of purported psychic phenomena. Some have an interface, or a point or a surface of contact, with the material world—thus the case of telekinesis. Other kinds of psychic phenomena appear to take place entirely within the domain of the psyche—one mind affecting another one—and then the only material trace is in the time and place of the alleged occurrence. Any attempt at explanation then attaches to the very fact of consciousness that evades most of the usual criteria of objective evidence.

PART OF THIS WORLD?

There are claims to psychic phenomena that do not exclude the possibility of some physical base, such as an energy field. Such is the case of Kirlian photography, which supposedly is able to photograph a person's aura, interpreted as an electric field surrounding a person's body (Krippner 1980). The same author also claims a material, not a transcendental, explanation for hypnosis, a well-known set of facts for which no rational explanation has been offered. In some cases it is claimed that hypnosis can be induced at will, but only by certain persons who are said to be gifted in that direction. What may make them equipped in this way is as unexplained as many other facets of our mental endowment.

The basic difficulty with hypnosis is the same as that of gravity: action at some distance, sometimes considerable distance. It is easy, in the present climate of scientoid culture, to assume that there must be some kind of energy progressing between the hypnotist and his object. There is not even the beginning of a theory as to what kind of energy this might be—it is just assumed, because action at a distance is assumed to always depend on some material influence going from here to there. The parallel with gravity is hardly reassuring, for the hypothetical "gravitons" assumed to energize it have not yet been found.

The wider category in which hypnosis belongs might be that of mental telepathy—much talked about but not much explored. In its simplest form, we all know about "eye contact" as a way of communicating a great deal—by inference or by other means? No one has shown how. Eye contact with animals only underlines the importance of this kind of telepathy.

The same problem extends to telekinesis (or, psychokinesis) and extrasensory perception. On both of these there is a considerable body of empirical data, but as yet no explanation (Irwin 1989). In the case of psychokinesis (poltergeist), Irwin found that most cases can in some way be traced to the apparent victim as the cause of the occurrence—but not all cases. In any event, the difference against a natural phenomenon such as gravity is that it has not been possible to establish firm rules under which these psychic phenomena may occur. We cannot always produce these phenomena at will, even if some progress has been made even in that direction but assuming (without further explanation) that some individuals are more gifted for (or prone to) psychic phenomena than others. The parallel with individual aesthetic and religious gifts, as discussed in the preceding essay, is intriguing.

Irwin clearly states his case for continuing research efforts: Whatever the explanations may turn out to be, the phenomena as such merit attention because they occupy some space in some people's lives, and it is, therefore, of interest to know more about them, whatever this may lead to. Such a basically agnostic attitude is, of course, much more acceptable in the world of science than the naive scientism that wants to reject anything that cannot now be explained.

Here is not the place to propose answers to these vexing problems. By way of example we might cite Michael Talbot's hypothesis about holograms in important areas of reality, including our brains. In this connection Talbot cites a conversation he had with his very good friend Whitley Strieber, suggesting that Strieber's "visitors" were not real but projections of Strieber's own psyche. The brain as hologram might hold all sorts of "alternative" views of reality in most of us without interfering with real business.

On the borderline of what might be part of this world, as it is usually understood, are the cases of "psychic archeology" and "metaphysical knowledge." The former is presented with an impressive collection of case histories where mediums appear to have helped or accelerated the discovery of archeological remains (Goodman 1977). The names of Frederick Bligh Bond in England and Aron Abrahamson in America figure prominently among these cases.

The problem is similar to that of the use of psychics or mediums in criminal investigations where some positive results are reported, although without criteria to make them repeatable at will. The parallel with special, individual gifts for beauty or religious experience comes to mind again. How such insights are obtained remains mysterious and hard to ideate. The facts nonetheless seem stubborn, and the end result—new knowledge about facts belonging in time, space, and quantum—suggests some relation to the conventional world.

Still in the same category of possibly relating in some yet unknown way to the conventional world belongs the general area "metaphysical knowledge." Foremost are the "life readings" of Edgar Cayce (Bro 1989). Such experiences relate to the conventional world, and more effort might be spent on verification. The same is harder to suggest in modern revivals of ghost appearances in visible form (Norman 1987).

We might pause and contemplate how an eagerness to disprove sometimes ends up in contradictions as bad as those to be dispelled. An early case was attempted by Bulwer Lytton, a prominent literary figure in the early 1800s and a friend of Disraeli. He built a hypothetical ghost story wherein a variety of upsetting psychic disturbances all could be traced to an ingenious electrical gadget that was concealed in a closed vault in the building. To explain why someone had entertained the notion of scaring his fellow mortals, Bulwer in the end comes up with a perpetrating personality that in some ways is as supernatural as the ghosts he wanted to displace. Ian Stevenson (1974) has a modern parallel when discussing xenoglossy: Those rationalists who cannot get rid of the evidence for this unusual language ability advance instead the hypothesis of "Super ESP" (extrasensory perception of exceptional quality). Stevenson's point is well taken: Super ESP has not been proposed in any other connection; it is called to life only when the apparent need comes up to deny something else that may be even more difficult to accept. Prove too much, and you prove nothing.

OUT OF THIS WORLD?

Less tied to verifiable consequences are the purported psychic phenomena of out-of-body experiences (OBEs), near-death experiences

(NDEs), and purported prenatal memories suggesting reincarnation. There is an intriguing gray area in xenoglossy (a medium talking in a language that the medium does not know).

Out-of-body experiences are a rather free-standing category in which verification will hinge upon someone obtaining some knowledge that could not have been obtained in any other way (Monroe 1971, 1994). It is often difficult to rule out some forgotten memory that may have been reactivated by a dream interpreted as an OBE. The rather clear-cut distinctions between several different types of OBEs, as sometimes reported, are in any event hard to explain away (Moody and Perry 1990, 1993). One suggestion made in this connection is that OBEs might represent some kind of mental hologram (Talbot 1991). Many attempts have been made to explain away the whole category (Blackmore 1993).

Near-death experiences have been reported for years, but they have been studied systematically in recent years, beginning with a ground-breaking book by Raymond Moody (1975) (Atwater 1994). Among the more intriguing findings is the connection of NDEs with a specific area in the brain, the right frontal lobe, which is therefore sometimes designated as "the circuit board of mysticism" (Morse and Perry 1990, 1992; Blackmore 1993). Here the NDE appears to blend with the many facets of brain-mind connection that have not yet been explained, but are accepted as fact. Morse and Perry retain the notion of NDE as something else—or at least something more—than a mere physiological brain function. The very link between such functions and mental experience has, in fact, not been explained in full. Morse and Perry add that, whatever explanation is eventually found, the NDE phenomena are worth exploring if for no other reason than their impact on people's lives. Those who have been through an NDE often are changed in their personalities, usually in the direction of having less fear. Other writers have taken the brain connection as proof that NDE's are nothing transcendental at all, just another physiological reaction. They would be hard put to explain how such a brain function might have come about in the first place—waking up with less fear of death is certainly no evolutionary advantage in a natural world where fear of death is necessary for survival.

NDEs contain a suggestion—but just barely—that the human mind or soul might continue to exist after death of the body. At the other end of the same spectrum are the reports about prenatal memories, suggesting reincarnation of some deceased person's soul into a newly born body.

The striking cases are those of purported memories, in most cases by young children, about a previous life. In several cases, these memories appear to be corroborated by checking with the environment of the purported previous life (Stevenson 1975; Banerjee and Oursler 1974; Bache 1991). There are obvious problems of verification, especially where the memories were checked only after some time had elapsed. The compilers of such case histories consider some of them to be "strongly suggestive" of reincarnation, but even these writers stop short of claiming definitive proof. The objection sometimes made that these more or less verifiable cases all relate to people in prominent or wealthy circumstances does not have to have any weight against the evidence as such; it may mean nothing more than a tendency for families to pay more attention where socially attractive connections are on the horizon and to disregard those of possible poor relations.

With such findings in the background, it also becomes harder to dismiss all results of regressive hypnosis as being all due to cryptomnesia as if often claimed (Baker 1992). The results reported by Whitton and Fisher (1986) and Weiss (1996) contain some cases that come close to historical verification. Such findings also render it more difficult to dismiss all recent reports about ghost sightings (Norman 1987).

The technique of hypnotic regression went out on a weaker limb when it was used for prediction of future events. It is intuitively less convincing that such should be possible because the events must then be understood as preordained, a condition rare in a world full of chaotic conditions. Such predictions are also up for verification, some of them rather soon (thus, those reported by Snow 1989).

Between near-death experiences and purported reincarnation memories lies the suggestion that the departed soul somehow exists after death and before rebirth. Ghost stories are not very convincing here, but xenoglossy might be more so. Of three cases reported by Ian Stevenson (in Swedish, German, and Bengali), at least the Swedish case (Stevenson 1974) contains many realistic details pertaining to a set locality that in this instance must have been completely unknown both to the medium and to those who conducted the interview with the purported Swedish-speaking spirit. Two Swedes who helped in the interview were apparently not familiar with the conditions of that part of the country. If a few cases could be found of similar caliber, we might one day state with confidence that departed souls have in fact continued to exist—after some fashion—beyond the pale of death. That may still be a long way from explaining how and why.

ANTHROPIC PRINCIPLES

There is no way of evading the general principle underlying the anthropic ones: To investigate something, we must do it with instruments at our command. We must also do it in ways that do not destroy ourselves in the process. We may have to beware of more than physical destruction: If some of the scientific-technological enterprise tend to damage or destroy our humanity, more may be at stake than just abstract truth.

Anthropic principles are attributed to Brandon Carter (1974, as cited by Casti 1989). Four such principles have been proposed: the weak, the strong, the final, and the participatory. They represent different assumptions as to what our minds can and cannot do. Further flights of fancy along similar lines have led over into "biological Copernicanism" (Seielstad 1989).

The weak anthropic principle (WAP) simply states that all our observations are limited by the necessity that they must be compatible with our existence as observers. As observers this principle does not seem to place any limits on our theorizing about things no one has ever observed, such as the Big Bang. Are we really to place more credence in the outcomes of mathematical extrapolation than in those of direct observation?

The strong anthropic principle (SAP) says that the Cosmos must be as we know it—else life could not exist. This merely restates the basic fact that our cosmos must be such that life could come into existence (Lightman 1991). It does not really confirm that all our recently received wisdom is true, nor does it say anything about the question of whether the Cosmos might have been significantly different from what we are familiar with. Theoretically, there might be a good deal of latitude for variant cosmoses to exist, but we do not know it. (Barrow 1988). A powerful reminder of the strictly unique character of our cosmos was delivered to us when it was found that the building block of heredity (the deoxyribonucleic acid, or DNA for short), including its forerunner in the RNA, is the same chemical compound in all living things on Earth, from archaebacteria to human beings, and including also the viruses that cause influenza or AIDS. DNA is a complex compound and (with RNA) it is the only one that has proven capable of serving as a continuing replicator of living tissue. The "toolbox of creation" (more on that in subsequent essays) is extremely restricted on this point. Whether some other chemical formula could have served as the basis for life's continuity in some different cosmos is, of course, entirely conjectural and not likely to be tested any time soon.

The final anthropic principle, or FAP (Penrose 1989), says that once life has been created, it will continue forever and mold the Cosmos to its will. This is, of course, just a credo, not a basis for science. Much will depend on just what is meant by saying that life has been created. It also depends on whether one believes in *panspermia*, the cosmic transfer of living spores from one celestial body to another one, as proposed among others by Svante Arrhenius (ca. 1900) and restated by Francis Crick (1981) and Fred Hoyle (1983).

That higher forms of life do not necessarily go on forever is confirmed by the several waves of species extinctions on our planet, sometimes affecting very large parts of all the biota of an epoch. A chilling reminder of the fragility of biota achievements comes from the Martian enigmas

(Carlotto 1991), which present the possibility that the red planet may have been the home of advanced life forms that are now totally extinct, maybe because of some ecological catastrophy ruining the atmosphere and the water supply (McSween 1993; Allegre 1992). Mars and Earth are the only planets in this solar system where life could exist, at least under the conditions the solar system functions at present. Both Mars and Earth are relatively close to being too cold or too hot, respectively, depending on the distance from the Sun. In order to continue forever, human life must first of all assume complete victory in the struggle to save the ecology of our planet (Bulloch 1989; Gribbin 1990; Asimov and Pohl 1991).

The participatory anthropic principle (PAP) says that the Cosmos is created by the collective observations of all intelligent observers. This apparently draws the ultimate consequence of a tendency of many scientists to say that reality becomes real only when it is observed (Casti 1989, Ch. 7). In this extreme formulation, the PAP certainly deserves to be rejected. In a less assertive way, the basic line of thought in the participatory anthropic principle calls to mind Carl Jung's suggestion that individual mental life, when lived successfully, may enrich the mind of God—certainly an inspiring article of faith for those who can believe in it (Jung 1938).

On the same assertive line of thought as in the proposed PAP are some of the bolder excursions into all kinds of planned space travel in the foreseeable future (Parker 1991; Halpern 1992; Gribbin 1992), with talk of eventual "supercivilizations," whatever their superiority may consist of.

At the end of the line is the sometimes occurring attempt at imagining the total void: Suppose there never existed anything, which leads the probing mind to a complete stop. The total void is unthinkable; we can never get out of the old truth of "cogito, ergo sum." The total void would never have had anyone to think of it. Even in the most poetic varieties, notions like "life is a dream" (Calderón) or "all is a fairy tale" (Hauff) assume the existence of some dreamer, or some storyteller, so this only removes the problem one step away.

INCOMPATIBLE IGNORANCE?

At this juncture, we are forced to pause and wonder, is it an impossible dream that we could learn to live with ignorance? And if we cannot do that, what would remain of the central Socratic message, "Know yourself"?

The depth of many people's craving for certainties is perhaps best illustrated by the still widespread belief in astrology. To the scientific mind, astrology is at the center of one of the few total certainties that we have: The certainty that astrology is pure and unadulterated bunk. Yet belief in astrology is so widespread, and so difficult to dispel by rational argument, that the conclusion is clear: The basic fact is a craving for certainties, a craving that is often stronger than that for truth. The same applies very much to the Bible-based so-called creationism (Eve and Harrold 1991) as well as to many cults, for instance the Urantia faith (Gardner 1995).

It is easy for the professional scientist to condemn other people's yearning for a faith to cling to, right up to the threshold of death. The scientoid mind tends to forget that scientism is itself a faith, based ultimately on assumptions that are not scientific. Like the illusions of political faiths such as Communism, the apparent salvation of scientism extends only to those who are active in formulating and propagating the faith. Their salvation lies in their ministry. Many ministers do not understand that the common parishioners—the rest of us, in the street—are not similarly supported by the very exercise of ministry.

When it comes down to ultimate foundations, we find that faith, hope, and love all defy the criteria of objective science. Yet we have no reason to believe that anyone could live entirely without them.

Could we live without any belief in received wisdom about the ultimate realities of our lives? Science as salvation was probed recently, with penetrating light (Midgley 1992). Living without any faith would be somewhat like living without beauty. Just try it (other than through flight into hallucinogenic drugs or the intoxication of workaholism) and see if it works (McKenna 1992). Programmatic atheists try to gain salvation from professing their dogmatic atheism—the least rational of all established faiths. But beyond the initial messianic stage, such declarations of negative faith fail to support anyone. The mysticism of mathematics is not much better (Barrow 1992).

Living without love is even harder to imagine (Lear 1990). Scratch the extreme materialist and you find love for money or power. In nature, love is pervasive, and if some people think they can live without it, they are usually badly off the mark.

Hope remains everlasting. For those who cannot hope for life beyond the pale of death, there is always the advice of Voltaire after the Lisbon earthquake had shattered his early optimism: Let us cultivate our gardens.

FIVE

Design: The Ways Things Are Made

Materiam superabat opus
(Workmanship excelled over material)

Ovidius, *Phaëton*, verse 5.

All reality has some features, some design to it. We will argue that without any design there can be no reality. Let us proceed step by step, beginning with a simple instance out of daily life.

THE ARCHITECT'S DRAWING

You approach a construction site—not, apparently, the locus of any great intellectual mystery. Excavators are at work, trucks are bringing in loads of bricks, cement, steel bars, and other materials essential for the building (or buildings) to be erected. Will it be a cathedral or a jailhouse? A high-rise apartment complex? Or maybe a line of one-family dwellings? The bricks and the cement give no clue; the excavating machines do not really either. The character of what is built depends instead on a non-material thing: the drawing by which the architect gives direction to the workers and by which he decides in advance how the building will look and what it can be used for.

The drawing is not a material thing. It does not consist of bricks and mortar; there is no steel or glass in it. If it happens to be drawn on paper, or reproduced on photographic film, or transmitted by teleprinter, all this is beside the point. A medieval architect could work without paper; he might have it all in his head. One way or the other, material paraphernalia of drawings and blueprints are unessential. Like the bricks and the mortar, the paper and the film could carry any number of different drawings. The essential property of the drawing is instead an abstraction or a set of thoughts: proportions, relations between parts of the whole, shapes of straight and curved lines, proportions between the materials used, and so on. All of this, this abstract design, "rules over" the use of materials. It is in itself not made of any physical material. It is pure thought. Thoughts may need a brain to be called to life, but once thought out, they can be (and often are) ideated independently of who first thought them out.

The power of this notion becomes evident when we try to remove it. Without adequate design, the new building would collapse. When this occasionally happens in real life—because either the architect or the construction engineer was incompetent or corrupt—the resulting heap of rubble still includes all the materials that had been used for constructing the faulty building. Not a single brick has been lost, and if some glass was broken, the fragments still add up to as much glass as there was before. As far as anyone knows, not one atom was lost because the building was reduced to chaos. When demolition comes by way of the wrecker's ball, the energy quantum spent in operating such equipment bears no particular relation to the design of the demolished building. It may cost as much to wreck a jailhouse as a cathedral.

There is more. The power of the drawing to compel the organization of bricks and mortar into construction according to its specifications need not be exhausted in one building site. If there is enough demand, the drawing may be used again and again, causing many buildings to be "identical" as far as their own appearance and functionality goes. We do, of course, recognize differences due to setting or location, which in various ways can affect both appearance and functionality. But setting can itself be taken as a design feature, as any landscape architect will be eager to point out. His drawing is analogous to that of the architect. The features of an untouched landscape also embody an implicit design that can be adopted for use or rejected.

The architect's drawing is, in fact, more essential than the bricks and the mortar, for any building may be erected from a diversity of alternative materials. Precast concrete slabs, for instance, can still give a building the same look and the same functionality as with bricks and mortar.

Let us rub it in: The architect's drawing is an essential part of reality. It is not made of any material, but it decides how materials are used. It is a product of thought. It is a set of thoughts.

Extend this by looking out over a city from an upper floor in a tall building. All that we see there has been dug up out of the earth, some place or other. Clay pits, gravel pits, lime quarries, and other scars from our extractive industries are negative design features inflicted on the face of the earth in order to gain the stuff that is required to build cities. More than architects' drawings went into this transformation of matter, but all can be understood as design features on one level or another. Brick factories, steel mills, all the manufacturing and servicing outfits that processed the materials and hauled them to the construction sites also contributed design features at various levels of materials treatment and transformation. They did so by applying the designs that were embodied in their own construction, ultimately the inventions that had to be made before such industries could be built. Thus, the clay was made into bricks, limestone into cement, iron ore into steel, and so on.

We need not argue here about the amount of energy that was used in these various processes. Energy is used also in the construction of buildings, but there is no reason why the energy needed to build a cathedral should be any different from that spent in erecting a jailhouse or an apartment complex. Neither is there any reason why more energy or less energy should be spent on one building material or another. Whether brick making requires more energy than the manufacturing of concrete slabs is beside the point; it has nothing to do with the appearance or the functionality of buildings. It remains true that the city rose out of the clay pits, the quarries, and the mines because of some system of thought. State of the art materials processing, engineering in the way materials are combined into viable structures, architects' and city planners' drawings of buildings to be built and urban landscapes to be created—all this needed thought along every step of the way. The pueblos of New Mexico, Nebuchadnezzar's Babylon, India's Taj Mahal, Peking's Gate of Heavenly Peace, and Manhattan's skyscrapers have all had their materials dug out of the earth and transformed by the state of the art, the architecture, and the economic and aesthetic ideas at the disposal of the builders and their rulers. The differences between the many buildings and urban areas come not from the earth as such but from how it was manipulated. People took the clay and the lime and gave them new designs. Embodied in solid matter, the designs appear both "real" and "material." To emphasize the point further: The design is not made of any clay, but of thought.

Architecture was chosen here for single-minded exposition because it is easy to visualize. In the same manner, the reflections made above apply to all human artifacts, and to some maybe even more strikingly upon second reflection. Miniaturization of tools and instruments is a powerful reminder of the supremacy of design over matter. A small computer may compute as well as a large one, and a wrist watch is more useful than a grandfather clock. In many cases, size is in fact irrelevant: What a time piece does is measure time, a function that has very little to do with the physical size or weight of the mechanism that does the work. Microdots can store information as accurately as large-type books; the information is the same.

In human artifacts, matter takes a back seat to design. How about the products of nature? Let us look first at organic matter and reflect on how it is built.

ORGANIC MATTER AND THE GENETIC CODE

You can draw a picture of a human being, or an animal, or a plant, but only as a reflection of exterior appearance. The built-in design that makes a living being we cannot yet draw or visualize in full. Much less can we do that in regard to the specifics of a given species, variety, or individual specimen. Unlike the building of cities and the making of other artifacts, we can call living beings into existence only by manipulating nature's designs. We cannot replace them from scratch—at least not yet.

This is worth pondering, for the designs of biological nature are more complex and more complicated than anything yet invented or contemplated by our Promethean spirits. Frankenstein's monster is as yet but a figment of imagination. Crops and livestock can be modified by various breeding tricks into "cultivars" and even into "cultigens." These terms mean varieties and species that are brought into existence because of interference from human culture at some stage or another and whether such interference was intentional or not. Some biological innovations have, in fact, been accomplished by trial and error through many pre-scientific millennia. Mankind has not only played rough with the face of the Earth, but has also greatly modified the plant cover and the animal population that occupy its biosphere. Wheat and corn, sugar beets and strawberries, dogs and cattle—all bear some imprint of human activity modifying nature. Vast natural ambients have also been changed, as from forest to steppe and from steppe to desert. Some of the prehistoric changes in crops and livestock are linked with such man-induced changes of the ecology.

When manipulating nature, mankind may even have modified itself by unconscious massive selection. Homo sapiens may well be a "cultigen," a culture-induced species, lacking any option of "reverting to nature."

Regarding crops and livestock, science has added many fresh details and is likely to continue doing so at an accelerating pace. Yet all of this, so far, is manipulative. Creating complex organic compounds such as proteins "from scratch" out of inorganic matter has not been practical to date.

The simplest known organic protein molecule consists of atoms estimated in the thousands. Such compounds are not just "polymers" or multiples of simpler ones. They often have a complex inner design. The same quantities of single elements, and in the same proportions, too, may turn out to be quite different compounds often with drastically different chemical properties and biological functions. This is so not only in organic chemistry. For instance, the CN molecule (one carbon atom, one nitrogen atom), which is the core of all the highly toxic cyanide compounds, contains the same elements that, in different compounds, are both beneficial and necessary to organic life.

Carbon, nitrogen, and so forth—all the simple elements are the same in a human being and in an earthworm, in a cow and in pasture grass. The differences between all the manifold organisms do not stem from those of the basic elements, but from the ways in which these elements have been combined. Like the architect's drawing, the designs of organic chemistry rule over matter that can be ingested as food and transformed to meet the specifications of the ingesting organism. These specifications—the organic designs—are themselves not made of carbon or nitrogen or any other chemical element. Like the architect's drawing, these specifications of organic designs can be read as ideas or abstractions. Their essence is in proportions and combinations. Like architecture, they can be reproduced full scale or reduced scale. Nature often miniaturizes its own creatures even without losing any essential functions.

All of this has come into startling relief through the advances of modern genetics. The basic fact of heredity has, of course, always been among life's recognized trivia, most of them "wonders dulled for us by repetition" (Robert Louis Stevenson). Hesiodos (ca. 700 B.C.), a not very profound Greek thinker, pointed to family resemblance as a sign of a virtuous society. North American Indians even divined the sexual nature of the corn plant, if with an erroneous explanation, but they did it long before European botanists had begun to grasp what pistils and stamens are doing in phanerogamic flowers. But even the early scientific attempts, by Lamarck, Darwin, and Mendel, at bringing some order into the subject of heredity could not yet fathom the immense complexity of organic chemistry in its commanding phase of genetic determination. We still do not know how this is done. Biologists have begun to identify defective genes, those causing hereditary disease, and how to repair them. But how the genes contrive to force the cells of an embryo to specialize into the various organs of a complex body still remains unknown (Shapiro 1991).

The building blocks of heredity are not only complex and complicated. The double helix inside these building blocks defies the imagination of most of us. These building blocks have the power of deciding in advance what manner of living being they will bring into existence. It is not as if they had any free will; each can only create what is laid down in its own code and does so only if no important obstacles intervene. But a germ cell commands matter to form an organism seemingly more complex than the genetic material from which it springs. This intriguing process has caused some scientists to compare the genetic code to a message. Its commandment might be translated as "give me x carbon atoms, y atoms of this and that, plus one sodium chloride molecule," and so forth. The orders are obeyed unless some elements happen to be missing in nutritive deficiency. Even then, a fetus can command some materials from the mother's body, shifting the burden of malnutrition as far as it can be shifted.

The concept of the genetic code as a command system has significance for our thinking about the coming into existence of organic life. For now, we should only dwell on the same essential lesson of the construction site and the city: The atoms by themselves do not make life. The design of organic matter is as indispensable as the chemical building blocks that it organizes.

There is more: An organism once brought into existence can, to a large extent, exchange the material of which it is built, little by little, through continuous organic processes. Within certain intervals, we have "a new body" (or most of one, anyway) without this in any manner having to change the character of the organism. A house could also be rebuilt, a few bricks at a time, without the architecture being in the least modified.

So the body is identified by its design, not by individual particles in its physical substance. The point comes forth starkly when an organism dies and its remains are destroyed. "Ashes to ashes, dust to dust," is a powerful symbol of destruction that does not reflect any loss of matter. Not one single atom has been lost; they are all there in the ashes, the dust, and the smoke from the pyre. Only the complex design into which atoms were arrayed has been destroyed, and that even before the dead body was committed to its final destination. The world's most impressive creatures are found to consist of proportions and arrangements: designs ruling over simple matter.

These proportions and arrangements—the designs of biological nature—can, like the architect's drawing, be thought of as capable of being recreated when the right conditions exist. Back-crossing from modern crops has sometimes succeeded in reconstructing the genetic combinations of more primitive crop strains that were extinct until then. If ever organic chemistry comes to the point of being able to build genes out of simpler matter, all extinct plant and animal species could conceivably be created anew and then some—but most of what a laboratory scientist could dream up would not be viable in the world we know. The potentially viable ones (whether they once existed or not) are likely to be a mere fraction of all those theoretically possible. On some distant planet, with geophysics and chemistry similar to those on our earth, there could conceivably come into being biological creatures identical with or closely similar to those we know, but also with variants adding up to a very different biological universe.

Each known species has among its properties the property of being viable, not prone to self-destruct or to be nonfeasible in the existing environment. Once viable in one environment it can, in most cases, adapt and remain viable in any reasonably similar environment. The viable species, in a sense, is an idea that might have been thought of beforehand, complicated as this would be to us and to anyone we know personally. Like the building designed by an incompetent architect, the nonviable organism would soon be eliminated, still without any loss of the constituent atoms.

Organic chemistry, with its plethora of complex designs, gives ample illustration to the general principle that the properties of a given phenomenon cannot be derived merely from studying its component parts. Design rules over matter; the atoms that give life in one configuration can be poison in another one. In organic chemistry, the principle of synergy is often very important: Two or more components of a complex compound may together have effects that are qualitatively different from the sum of those of the components. We shall have more to say about synergy in the next essay.

CONFIGURATION OF ATOMS

So far we have contemplated the configurations that make bricks into buildings; clay into bricks; atoms into clay and into carbohydrates; and proteins and double-helix genetic determinants ultimately into living beings, sexes, and individuals. We could use the simile that the natural world with its natural substances, its inorganic and organic compounds, and its living plants and animals is all an exceedingly complex architectural design arraying matter this way and that way. Coming down to matter and energy is where classical materialism built the edifice of its doctrine. Matter and energy are indestructible—they are what reality is made of—so the classical materialists believed—and many people still seem to believe so. This simplistic notion appeared logical as long as the several forms of matter—the chemical elements—were thought of as immutable.

In the world of immutable chemical elements, carbon was carbon and nitrogen was nitrogen, each with their physical and chemical properties settled once and for all. The *atom*, the indivisibility of ancient Greek thinking, was the ultimate carrier of existence. When each chemical element was thought of as immutable, all characteristics of real things must draw ultimately on the characteristics of the primary chemical elements.

This solid ground of classical materialism has crumbled as a result of nuclear physics. No longer are the elements immutable or impossible to reduce one into another or all of them into simpler components. Even the borderline between matter and energy has been dissolved. An atom is a bundle of energy quanta, wrapped up in a stable configuration. Energy (or, at least, electromagnetic energy) is quanta on the loose, not wrapped up in stable configurations, but still characterized by a number of alternative design features in wavelength and spin. Maybe there are still other distinctions to be made. By some physical process, in a laboratory or inside a star, matter can disintegrate into energy quanta, eventually the simplest form of electromagnetic energy that is assumed to be light particles or photons—we will have to leave out the elusive "neutrinos" until more information comes from the physicists (cf. Mann 1997, pp. 24, 32). From the cataclysmic storms of the Sun, fragments of its matter flow out into empty or near-empty space as starlight—and whatever else.

Thus, even the atoms reduce their substance into design: Carbon is carbon and nitrogen is nitrogen, not because they are made of different kinds of building blocks—they are all made of the same kind of elemental quanta. Their different chemical properties come instead from the weight of the quanta that are bundled up in their stable configurations and from the configurations themselves: size of the nucleus, number and character of the particles that are kept in balance within and around the nucleus, and shape of the surrounding "electron shell." In each case, some design has made elemental particles form one chemical element or another, not at random but according to which designs are viable. That is, they can remain stable for some time or indefinitely, so long as no superior force intervenes to break up the design. Some of the elements in the now rather complete periodic table may not be fully viable, such as astatine and technetium and the heavier among the transuranians.

The general principle can be well-illustrated from carbon and nitrogen, which have isotopes with the same atomic weight. The difference between carbon-14 and nitrogen-14 is slight as far as the weight of particles goes, but the chemical difference is striking. The fact that the transition from nitrogen-14 (regular nitrogen) to carbon-14 (radio-carbon) appears to take place in the atmosphere and appears to be due to the intervention of vagrant neutrons only underlines how essential the design is. When carbon and nitrogen also can act together as basic materials for biological tissues (proteins) or as deadly poisons (cyanide), we again get the same lesson driven home: Design rules over basic substance.

It is, of course, true that nuclear designs are the result of processes that are entirely different from those of a construction site or a reproducing organism, but such differences are beside the point here. The point is, rather, that whatever created the carbon atoms and the nitrogen atoms—and the other elements—did in fact establish complex designs in which the decisive thing, for the characteristics of what was created, was not the identity or even the prior characteristics of the quanta that came to serve as building blocks in these atoms. The decisive thing, for the kind of reality that was created, was the design that was made to rule over the simple building blocks.

The primary significance of design over basic stuff is reflected also in the fact that all matter of which we have direct experience is quite porous. What fills space is not so much the quanta that are wrapped up as atoms and molecules and so on, but the designs that hold the quanta in place. Lines of tension between atoms hold their parts in place and prevent, up to a point, other objects from penetrating them. The same kinds of influences also prevent matter in all its habitual forms from being compressed below certain densities. Not only do solid objects resist compression, but so do liquids and gases, witness the effects of hydraulic pressure and of gas explosions. From advanced research it now appears that if quanta (energy/matter) could be compressed into structureless amorphousness, then all the stuff we know about might occupy much less space than it now occupies. Inside a "black hole" in some peculiar corner of the cosmos (if such exist), Earth might not take up more space than an orange. A whole galaxy might not be more space filling than Earth is now.

This view of designless matter brings us down to energy (electromagnetic energy) as the apparently ultimate, "rock bottom" phase of material reality.

ENERGY: FORM AND DENSITY

Since all matter can ultimately be reduced to energy quanta, we are next led to ask: Does reality derive any of its qualities or characteristics from "pure" energy? This has often been believed in quasi-romantic speculations about soul versus matter. It has also been a tenet in some late phases of materialist thinking.

Does energy in all its forms have inherent qualities that decide what objects or processes can come into existence because of these characteristics in the building blocks of elemental energy? If this were so, it would form an apparent contrast against the entire previous analysis that proceeded from the architect's drawing through the gamut of complex designs in chemical elements, inorganic and organic compounds, and biological beings. All of these, we have argued, derive their qualities from design features imposed on building blocks that are in themselves essentially neutral to the use made of them.

The question, like the proverbial horizon, seems to recede when we try to approach it. The question of the *form* of energy appeared to be answered from our attempts at energy economics. The expression "how much energy we consume" turned out to be misstated. When energy is defined as heat units, we do not really consume any. The first law of thermodynamics states that energy is indestructible. This may, in the end, be no more than a logical tautology; but in the circumstances that prevail in our terrestrial, or immediately interstellar, environment it apparently holds with a vengeance. When units are indestructible they are also, in their purity, of hardly any value or consequence. What is of potential value or consequence is instead certain forms of embodied energy that are available for use. The breaking down of such forms releases heat or other energy for economic use or for other consequence. Hydrocarbons, fissionable or fusionable nuclei, position energy in water basins, kinetic energy in wind, falling water, and so forth—we can consume them because they are destructible.

In the more precise formulations of energy economics we consume fuel, not energy. We do not consume any heat units. What we consume is instead the form in which we find the energy quanta pent up—a design, analogous to the other designs we have already discussed. There are, of course, also energy quanta pent up in all sorts of objects all around us, but mostly in forms that we cannot exploit economically, nor could they be released for other consequences than those of our economic life. There is also the atmosphere's ambient heat, continually replenished from fresh sunshine and also continually depleted by solar heat escaping back into space; this is an enormous flow that we can tap only with great difficulty, but is of decisive consequence for the quality of the planet's ecosphere. Fusion of nuclei in elements such as deuterium and tritium (and maybe some other light elements) is theoretically possible but as yet eludes us economically, while playing the main role in the processes inside the stars. Out of reach, economically, are both the electricity of thunderstorms and the winds of hurricanes—again, both of which have great consequences for the workings of nature on our planet.

In short, what makes an energy source for our purposes (or for any purpose of doing work) is the *form* of the energy, including its density. Electricity can be used in more ways than heat or light. Chemical compounds such as hydrocarbons or nuclear structures that are fissionable within our technological reach we can use for economic effect. Even some simple elements can serve as energy sources if they oxidize at temperatures we can handle. Elemental sulfur is actually used as a heat source in fertilizer factories. Zinc, for instance, can burn at a few hundred degrees, leaving zinc oxide as a residue; only elemental zinc is too expensive to use in this way. With variations, the same goes for nature's household: What makes a substance or a situation a source of available energy for the purpose of natural processes is again the form in which the energy is present or embodied. Even the Sun does not seem to burn all its elements in the same phase of solar existence.

Logically, we then conclude that the work done when energy is released by chemical or nuclear (or other) processes is due not merely to the quanta concerned, for these, in some circumstances, can be quite ineffectual. Rather, or even in the first place, the work done depends on the design in which pent-up quanta are held in place before they are released. Power is in form (including density) rather than in mere quantity.

Now the second law of thermodynamics states that all energy processes are predominantly downhill. They go from higher (more complex and more efficient) energy forms toward lower (simpler, less efficient) ones. This would lead to the often debated consequence that the Cosmos would eventually end up in "heat death," a state in which all free energy would have been degraded into heat units dispersed at low density. This apparently leaves out the quanta still embodied in stable matter. Because of the immensity of interstellar and intergalactic space, the ultimate dispersal of all free energy as low temperature heat would lead to such low densities that these units become virtually inert, incapable of setting off any consequences whether labelled as work or otherwise. Stars are spaced so thinly through space that even their cauldrons of millions of degrees can raise the temperature of the great void only insignificantly above zero. At 4^k (four degrees Kelvin), no nuclear or chemical processes can take place.

THE "ENTROPY LAW"

Starting from the second law of thermodynamics, first formulated by Helmholtz (1847), the so-called entropy law as an ultimate limit on all human endeavors has been much elaborated upon (Rifkin and Howard 1980). Relating this to human endeavors may confuse scales and magnitudes. What happens to the stellar cosmos over millions, nay, billions of years is not necessarily relevant for our human planning, which may concern centuries or, at most, millennia. Life on Earth may not be very efficient at utilizing incident sunlight, but then the supply is vast (by our scales of measurement) and continuously self-repeating, so the efficiency of converting sunlight is not necessarily of primary importance compared to what is generated by help of such conversions of energy.

How the equations of the energy laws may relate to the building up of more complex, qualitatively higher forms of existence is a different problem that cannot be resolved by mere reference to quantity measurements. Life derives its power not only from embodied energy quanta, but also from highly complex and efficient design, a thing not quantitative in itself. Efficiency of design will be discussed further in the next essay, about the processes of nature.

There may be a hidden fallacy also about the second law of thermodynamics. It may reflect conditions as they are found on Earth and in the astronomic universe as it was known at the time of Helmholtz. Recent theorizing about black holes may point to one possible origin of such a fallacy. If there are such centers of gravity that they cause uphill movement toward higher densities of both matter and energy, then they may eventually set off processes leading to more complex nuclear and chemical designs as well. The very formation of stars from diffuse clouds of gas in the Cosmos—if that is how stars are formed—would also represent a case of uphill movement. So also would the formation of all the heavier elements (beyond helium) inside stars where the prevailing extremely high temperatures allow continued fusion into heavier elements, from which planets are eventually formed.

Those black-hole centers—if they exist—apparently are invisible because they emit no light; even starlight is thought to collapse into their compacting force. As with radioactivity on Earth, our perception of how the Cosmos functions may be critically determined by the fact that to live we must, by advance definition, be in a livable ambient. In or near black holes there can be no life, not even any instrument of direct perception.

Power coming from the release of energy is, in any event, dependent on the design of the substance in which the energy was available. "Heat death" as the ultimate of material decay would be the extreme case of entropy, or the low point of featurelessness or lack of design. This would be an ambient where there is no active reality, the dispersed heat being too weak to set off any effect.

The ultimate residue of material decay is assumed to be a very small elementary unit, labelled *photon* in the parlance of modern physics—so small that it even is supposed to have zero mass. As starlight, the photons are traveling at the speed of light. Basically, all real objects in our world are thought of as built of immense amounts of elemental energy quanta, carrying power in relation to the frequency of their wave movement when free, sometimes also spin. These units are thought of as clustered, combined and architectured into various designs eventually of very high complexity. All designs can be destroyed and usually they are, sooner or later, leaving behind the elemental energy particles as the ultimate reality. Are they?

WAVES OR PARTICLES?

A standing controversy in physics is (or was, until recently) the seeming contradiction between classical and quantum theory: How can radiation (such as light) be both particles and waves? This seeming dualism has been seen as an analogy with the equally puzzling duality of body and soul, which has given pause to some physicists (Teller, Teller, and Talley 1991).

Seen from the perspective developed in this book, the seeming riddle appears more like a conundrum—a false problem. The physicists' attempts at tackling this kind of difficulty brings to mind the complicated "epicycles" by which some of the late Ptolemaians tried to rationalize the problems left unsolved by their earth-centric image of the Cosmos. To this lay observer it seems evident that radiation—such as light, heat, x-rays, gamma rays, and radio waves as well as electricity—should be understood as particles traveling at speeds that cannot exceed that of light, but in wavelike motion, with or without spin. The pattern of motion defines the kind of radiation that can be observed.

There is a recent theory of elemental particles that ideates these as elongated strings of something, maybe even with small protuberances making them resemble centipedes—if they could be photographed, which for obvious reasons they cannot. This theory may lead to the possible hypothesis that even these strings, as they move ahead, may be shaped in wavelike fashion, different for different kinds of radiation. This is not a necessary hypothesis, however. Even if the strings are thought of as straight, their moving in wavelike patterns may be reasonable.

The transition from one kind of radiation to another one (such as when light becomes heat) could mean that the wave function changes because of the very agent that causes the change in the type of radiation. In this case, a dark body absorbing light turns it into heat; the wave function would change because of the different medium through which the particles have to travel. The difference between red light and blue light, or between light rays and heat rays, can be thought of as analogous with the difference between cathedrals and jailhouses, or between apartment complexes and family dwellings. The photons without any wave patterns would then be comparable to bricks, while their wavelike patterns are designs, ruling over the elemental particles, causing them to have the specific effects they have.

If there were any photons without *any* specific wavelike pattern, they would have to be thought of as points or stringlike lines moving in a straight line. Such designless photons would probably also be completely inert, lacking any effect or impact on the matter through which they would glide effortlessly, being transformed again into specific radiation only to the extent that they collide with something coming their way. But maybe they, like light rays near the Sun, just bend around any meeting object, without ever colliding with anything. If this were so, we probably could not ever observe such featureless particles. They would be as much nothingness as empty space or time without events.

When contemplating such possible outer limits of positive existence, we want to avoid getting into anything resembling polemics over the intricacies of cosmological astrophysics. Things like the redshift of starlight from afar, or the pervasive cosmic background radiation, are tempting but not necessary for our main argument. The indestructibility of ultimate energy/matter in the Cosmos might be called in question by recent findings of dark matter in the Cosmos, or the Oort cloud of comets around our Sun (and many other stars too?) (Littman 1988). If there is long-distance attrition of traveling photons, this might cause a question mark also for the Big Bang theory, which is not without dispute to begin with (Lerner 1991; Lightman 1991).

QUANTUM, SPACE, AND TIME: REALITY WITHOUT SHAPE?

Trying to follow in our thoughts the last feeble residual of disintegrating matter, namely starlight, on its voyage through intergalactic space, we cannot escape asking the question: Does the starlight in the void still have any design to it? Would it still be energy—or matter—without any characteristics? Only density would remain, but it too might be entirely featureless. There might be no variations in extremely low density, as the escaping rays of starlight hurry away from their origins, spreading ever thinner with increasing distance from those origins, but out over ever wider expanses of near-void space (Davies and Gribbin 1992).

The question is whether energy quanta—the building blocks of physical nature—can exist at all if it has absolutely no characteristics, no design features? How could such shapeless existence be at all possible, let alone be understood?

To discuss this, we have to go into the slippery problem of infinity. It is one of the classical antinomies of philosophy that we cannot imagine infinity, yet we cannot imagine the world as finite either, for when contemplating the end of the world, we cannot hold back the intellectual itch of asking: What is outside the limit? Modern physics answers that what is outside is nothing. Time cannot exist without events, and space cannot exist without any content. The total void, not having even an outer limit, is not in existence. Both events and content must, by definition, have some design to them, else they would coincide with featureless entropy. If the astronomical cosmos expands, as it is widely believed to do, then it can do so without obstacle because the outside void would yield no resistance. By expanding, the stellar cosmos would create both space and time. The presence of energy/matter (quantum) creates finite distances, and energy processes create time frames where none existed before. Both distances and processes, in turn, require some design features; none of them could exist without design.

Now we can turn the usual reasoning inside out and ask: What happens to starlight when it leaves intergalactic space and proceeds into the total void, if that is what it does? Nothing happens, says the first law of thermodynamics, for energy is indestructible. Free starlight just goes on and on, without any limit in time or space. Would this not expand the Cosmos? Would the void still be void when it receives starlight from the Cosmos of stars and galaxies—black holes or no?

Here the first law of thermodynamics forsakes us, for it has been derived in our own safe corner of the Cosmos, which depends on the ultimate energy equation as much as on the stability of atoms and chemical compounds. As with the black hole, we also lack any instruments to tell us what becomes of starlight in the ultimate void. We cannot tell whether, in the absence of any other physical force than itself, the traveling starlight may have retained what little character it had after the downhill movement has been completed. In the void, starlight has to be inert when it cannot be reduced to any existence even more elemental.

Here we have the ultimate question: Is not elemental energy, quantum without design features, just another dimension of material existence, analogous to space and time? Is it not, like space and time, dependent on design to become real?

Here we can cite Heisenberg's uncertainty statement: Since an elementary particle cannot be observed with accuracy as to both position and velocity at the same time, it follows that we cannot say anything about either position or velocity without observing it. Observation always means interference: The photons must meet some other object that is either directly or indirectly part of the observing instrumentality. Left to itself, without any encounters of any kind, the traveling photon may well be inert, unreal. Its reality may be dormant and may be called back by collision with something else, something of one design or another, just as both space and time may be called into life by encountering objects or events. The reality of light striking our eyes, or a camera lens, or something else of whatever description—such as a rock where the light is turned into heat—that new reality coming out of the encounter then depends in part on the opposing reality.

We have argued previously that energy/matter (quantum) itself cannot exist without design features that are both spatial and temporal as well as quantum oriented. We now call into question whether quanta could be real without any design features. Matter is quantum wrapped up in stable configurations, such as atoms and molecules. Energy (more precisely, electromagnetic energy) is quantum on the loose, with design features such as wavelengths, sometimes also spin. This proposes a reversal of materialist doctrine: Even energy and matter can exist only at the mercy of some design features.

EXISTENCE AND COMPLEXITY

Since design is essential to reality, we conclude that more complex design must mean more intensive reality. In a qualitative sense, more design features mean "more reality." Proof is that as designs are destroyed, reality is diminished. This is true of a house on fire, a dying body, and a star burning itself out.

The agents that bring about more complex designs are many and varying and cannot be described in the same terms for all levels of existence. This may be one good place to point out—as we will, repeatedly—that the first axiom of logic, that "the whole is equal to the sum of its parts," is true only in a purely quantitative sense. In anything of complex design, the whole becomes something more and something else than the mere quantitative sum of its parts.

We will certainly abstain here from any attempt at speculating over the origin of matter in the Cosmos or in the stars. The subject has recently been made more puzzling by the news about "antimatter." Obviously the much discussed "heat death" has not yet occurred. We, therefore, tend to think of a beginning when the current downhill movement may have started. Maybe there was a beginning to matter as we know it, or maybe many beginnings, of whatever description they may have been.

Within the range of things we can observe it will be necessary to insist on increased complexity as a qualitative offset, even in reduced quantity, against losses of other complexity in the known energy transformation processes. Some increased complexity follows from the uphill movements in the universe of living things. The second law of thermodynamics may be quite true in the range of phenomena we know and can be concerned with, but this may not be all that can be said of energy conversions. If heat losses—that is, losses of complexity in matter embodying available energy—are incurred in order to obtain even higher complexity in a more limited framework such as that of living organisms, then reality as a whole may not have lost any of its intensity as a result. Such transformations toward higher complexity—that is, toward more intense reality—have a significance quite apart from thermodynamics.

Complexity versus mass: Let us use the example of "biological climax." Some ecologists use this expression to denote the state of affairs when an area carries the largest amount of biomass it can carry, given the soil, the climate, and other limiting factors. Biomass is defined as the dry weight of all living matter. When biomass is quantitatively maximized, the area is said to have attained biological climax. When less than maximum is maintained, this is termed "disclimax." An example would be when a forest is destroyed and replaced by grass cover or when forest is replaced by annual field crops. Human occupancy often leads to disclimax. Single-species stands, both of field crops and of forest trees, also often imply some disclimax in comparison with mixed stands, which are common in wild nature as well as in primitive agriculture. Mixed stands usually make more complete use of soil and water and sunshine than is possible with single species, hence mixed stands are more likely to attain biological climax.

The difficulty with such a classification of biological states is that it regards only quantity (without further description) and not the complexity that we argue is an index of quality or intensity of existence. The "climax" definition does not even consider the duration or durability of different kinds of living tissue. A food-grubbing Stone Age tribe might have destroyed a lot of forest, but their brains were a stratum of biomass of higher complexity and also of longer life span than any green or flowering plant or branch. A tree has been called "a society of organs," but the long life span of a tree should not be confused with the life span of each of its parts, which usually is much shorter. At the far upper end of human development, a city may, of course, be said to represent a rather high degree of disclimax, and so can the termite society, the anthill, and the bee or wasp colony. But by being the home of numerous human brains, a city may also exemplify a concentration of high-level existence if these brains are organized into a culture that generates art, knowledge, and power over nature. Throbbing with high-complex life, the "downtown" of a recent popular song may replace in quality what it ruins in quantity when mankind, the "ecological dominant," muscles out large amounts of the more elementary biological climax that would exist in a tract of wilderness—whether on the banks of the Nile, along the Yellow River or the Seine, or on Manhattan Island.

This, of course, is no defense of mindless population increase leading to congestion and deep, ingrained poverty. Nor does it defend political systems that exploit the human need of security to stifle creativity in individuals who possess that source of more intensive existence. There are certainly limits to the enriching effect of large population numbers. For complex culture, there has once been a lower limit in critical mass of population size, but that is long past and we should now be more concerned about the constraining effect of upper limits in the size of human population. There is more: If the physical disclimax of the "cultural landscape" is allowed to annihilate too much wilderness, replacing it with increasingly repetitive human accomplishments, then again some complexity is lost, resulting in less intensity of total existence.

Mankind may not be the only species that tends to constrain nature's complexity by biological success leading to larger numbers. The baboons of Africa are said to have been more numerous than the people of Africa not very long ago (Strum 1987). By their biological success they may well have limited the quantitative scope of other species and so brought about more sameness (less variety) already in the wilderness. Botany has many parallels in species that are more successful than others. Our purpose here is not to close a debate, but to open one up.

A QUESTION OF PRIORITY

The debates of classical materialism turned very much around the question of prior existence. Less complex things existed before the more complex ones; their prior existence was seen as a condition for the coming into existence of the more complex ones. Matter mattered more than spirit. The less complex things, acting as prior conditions for the more complex ones, were also thought of as causes—the usual fallacy of *post hoc ergo propter hoc.*

The real question should be: Can we think of things having certain characteristics—certain designs—without the designs being, in some sense, existing prior to the things in which the design is embodied? How else can the design rule over the constituent materials?

In one sector of reality, the answer is loud and clear: Artifacts, man-made things, can only come into being after a design has been thought out—ideated by someone. Without an architect's drawing of some kind there can be no building erected—certainly not if we expect the building to be viable, not collapsing immediately into rubble because the builder was incompetent. And so it goes with the gamut of artifacts, from log cabins to cathedrals, from knives to forks, from fifes to church organs, from Altamira's cave drawings to Michelangelo's *Pietà*. In the sphere of human artifacts, a design in someone's mind always precedes the forcing along of the material object.

How about nature's designs? We have no demonstrable knowledge of anyone out there who might ideate atoms before they were formed, or protoplasm before any such thing came into existence, or our brains' neocortex before it evolved. Can the underlying, or inherent, designs be thought of as, in some sense, existing before they could bring material facts into existence? If not, were they also created the first time any of them was embodied? Before attempting to tackle this immense problem, we must first think through the concept of processes of nature, usually called "laws of nature." As far as we can observe, these are immutable, and they are specific: Only those laws actually functioning are laws of nature. Many arbitrary formulations lack sense or consequences, leading, at best, to objects that are not viable, existing at most for a fleeting moment before they collapse as being not viable, like the building designed by an incompetent architect.

How many laws of nature are there and where do they come from? Our next essay will endeavor to sort out these problems and to propose some of the answers.

SIX

Process: The Ways Things Work

"πάντα·ρεί"
(All is flowing)

Herakleitos from Ephesos, ca. 500 B.C.

A seed sprouts into a plant. The plant develops leaves, then flower buds. Buds open into flowers. Flowers ripen into seeds. All this happens according to set designs, materializing by set processes. The example is from botany, but analogies run the gamut through all known reality, from particle physics to cultural anthropology. The world is as full of set processes as it is full of set designs. Like the designs of things, the designs of processes must be viable if they are to continue. Any nonviable process, like any nonviable design of a thing, soon falls by the wayside and ceases to be part of the reality we can perceive.

Even in the heyday of materialistic monism, the concepts of reality included an important facet that was neither matter nor energy. This facet is the law of nature. The concept had a prominent place already in the eighteenth-century Enlightenment. This is reflected in a phrase used in the American Declaration of Independence of 1776: "The Laws of Nature and of Nature's God." Linking the natural laws to divine authority, and vice versa, was meant to reinforce the respect for both.

LAWS OF NATURE

Viable processes are repetitive, hence their outcome is to some degree predictable. The laws of nature are invoked to explain why Cause A leads to Effect B. In modern science, the nexus between cause and effect is often understood as being due to the action of some "force." In truth, nobody has ever seen a force. What we can observe are merely invariant processes. Assuming these to reflect underlying forces may render the computation of practical problems somewhat easier. In some way, the interposition of a force between cause and effect represents circular logic. It also resembles the (usually rejected) notion of "élan vital" as an ultimate cause of life. Heinrich Hertz (1857–1894), the eminent physicist whose name lives on in the terms by which we designate the wavelengths of electromagnetic waves, tried to build a general theory of physics that would avoid the concept of force. Some analytical philosophers have blamed the force concept on subjective analogies with our internal, physiological processes as we experience them subjectively (Hägerström 1936). The equations of basic physics might have to be rewritten accordingly.

The laws of nature can be general or specific. The general natural laws are the bare bones within the complex anatomy of designs that constitute the realities of our world, as discussed in the previous essay. Recent reductionist thinking tries to derive all the laws from each other, eventually from a single, original force—really a kind of one-dimensional God.

When nature, and other reality, is studied in detail, it becomes evident that the specific laws of nature are many and often complex. The laws of nature, at one level or another, will go a long way to account for the descriptive features by which reality's designs can be comprehended. Recently, the concept of chaos has been called to renewed life as a counterpoise to the even more negative concept of entropy (Hall, [1991] (1992). Arbitrary disorder has been shown to include a range from simplistic to manifold disorder. The assumption of general laws as prevailing, and of specific laws as being always possible to derive from the general ones, owes much to the tendency of modern science to trust generalization from detail rather than the much more difficult perception of holistic phenomena.

Specific laws of nature can be cited from the analytical phases of all sciences. Physics, chemistry, and biology all contribute a vast catalog of natural invariances that we can call specific laws of nature. Analogous laws can also, at least in principle, be traced from human society in all its variety. When moving into this difficult area we must, of course, beware of many of the generalizations that are offered to us in the vast literatures of economics, sociology, psychology, and other disciplines treating some aspect of the human phenomenon. The limits of science in human affairs will be discussed in a later essay. It is probably significant that the new mathematical discipline of fractal geometry has large applications to several classes of natural phenomena, but none to human artifacts such as buildings or sculptures (McGuire 1991; Carlotto 1991; Mandelbrot cited in Hall [1991] 1992).

For now, let us concentrate on how laws of nature relate to the description and explanation of the designs of reality. Explanation, we shall argue time and again, always sooner or later boils down to a description of invariant features or processes. If we can portray the designs of reality as they exist and function, that may be all the explanation that can be attempted (Teller, Teller, and Talley 1991).

GENERAL LAWS OF PHYSICS

One trend in scientific thought that dies hard is the belief in general laws of nature that apply "everywhere and everywhen" (Barrow 1988, 1991). Such laws should represent general rules of conduct in nature, encompassing and restricting the movements of both matter and energy, and of the working or application of complex invariant designs that are (or can be) embodied in matter or energy.

Such generalities have, at various times, included the law of gravity, much celebrated since the times of Newton and Laplace, but recently under debate due to the theory of relativity. There seem to be unsolved problems as to how gravity can work at a distance unless there is a yet-undetected class of particles, the hypothetical "gravitons." There are also the laws of thermodynamics as we discussed in the previous essay, with some reservations as to their possible universal validity. Better explored is the law or the force of electrodynamics—rather, the laws that govern how electromagnetism works. In nuclear physics, much has also been written about the weak and the strong nuclear forces, governing, apparently, the cohesion of atoms and of the heavy particles (protons and neutrons) within the atoms. The breaking up of atoms by fission and fusion would illustrate the working of the weak force, while the strong one would belong in the much more difficult process of breaking up the heavy particles.

There is an important distinction to be made here between those putative forces that may in fact reflect design features in the things they are supposed to move and the laws or processes that decide how the various designs function.

Some of the frontier thinkers on physics are engaged in a tantalizing quest for a formulation that would prove to reflect a single unifying, steering force of nature, never to be broken or relativized by any marginal conditions. This search somehow brings to mind the ancient Greek satire about Zeus being dethroned and replaced by King Whirlwind. The single unifying force would in some sense be God, albeit in a very elemental and expressionless way.

A reason for this quest is in the observations that have led—by extrapolation—to the theory of the origin of the Cosmos (or, more properly, *our* cosmos) from a Big Bang, such and such time ago. Extrapolation from certain observation data has led to the conclusion that a primordial explosion has taken place from which our stellar and galactic cosmos took its origin. That explosion, apparently, is supposed to have proceeded from a starting body of extremely small size—maybe no larger than an atom. Faced with this astonishing result, some astrophysicists just assume that at this remote stage of the world's genesis, the laws of physics as they are now known to science would have to break down. Or, shall we say, they would not yet have come into play. Others try to explain how this might fit in with the general statement that there are laws of physics that are valid in all places and at all times. Cosmology may have reached an impasse (Lindley 1993).

The theory of the Big Bang is bold, and there are those who maintain that it never happened (Lerner 1991). As a modification, there is the possibility that this cosmos may not be the only one—there may have been (or there may still be) others like it at unspeakable distances in time and space and therefore totally out of reach for any observation or even the most daring theory. Even the Big Bang cosmos, if it took place, in its first stages would have been impossible to observe, even theoretically. At some early stage, there would have been a cloud of matter so dense that it could not have been observed by any instrument that current science feels able to build or even ideate.

WHAT IS "ENERGY"?

Contemplating the laws of thermodynamics, we meet a seeming contradiction: The second law, in some sense, contradicts the first one because not all energy processes are equally efficient. From energy economics we have the distinctions between fuel, power, and effect. Fuel (the energy source) has a certain embodied potential energy (enthalpy), not all of which can be recovered as power (energy actually deployed in the ensuing process). Not all power is used to full effect, the difference being dispersed as waste heat, hence the threat of eventual "heat death."

Einstein established the equivalence of matter and energy. In some cases, units of matter may be blown up into clouds of energy quanta. Matter is quanta wrapped up in stable configurations. Energy is quanta on the loose. The concept of waste heat becomes easy to visualize.

This fits with electromagnetic energy (electricity, light, gamma rays, etc.) because there the energy units are understood as actual fragments of matter having been blown up (or never having come together). But there are other cases where energy is not always present where it eventually shows up. Position energy and kinetic energy represent relations between some stable configurations and some configurations in a process of change.

Let us take a look at a coiled spring, for example. No energy quanta (in the electromagnetic sense) are present in its compressed substance. Some quanta may have been used up (reduced to low-efficiency state, waste heat) to compress the spring. Some energy will be released when the spring is allowed to uncoil. The compressing and the releasing are two matching processes; the energy used up and released, respectively, may be shown to be the same quantity. This equivalence does not have to mean that any physical energy quanta (like those of electromagnetism) were embodied or released, respectively. If the coiled spring had been destroyed by a hammer blow or by submerging into molten metal, then the energy stored in its coiled state may not be traceable, even theoretically. Maintaining faith in the equivalence is then just an accounting convenience, not an empirical result.

We may use the coiled spring as an example of energy processes outside the domain of electromagnetism. Anything that is created by an uphill process can be said to embody energy. The undoing of the result by a matching downhill movement can be said to release energy.

Thus, maybe, energy boils down to the difference between uphill and downhill processes. Energy in water behind a dam is not represented by any energetic photons present in place, and if the water thus stored all gets turned into water gas by dry weather, no equivalence to its position energy was obtained. The release of energy in falling water could be due to gravitons causing gravity to function, but it is by no means clear how this could relate to the energy released, which rather is a function between the weight of the falling liquid as related to the size of the attracting body, in this case the Earth. A meteorite falling on the Moon does not release as much energy as it does falling on Earth or on Jupiter.

In any event, the energy equivalences may hold only as long as we only look at quantities. For "qualia," qualitative distinctions, energy equivalence may not apply or even have any definable meaning.

PHASE TRANSITIONS

Why should the laws of physics as they are now known to science apply also in an early phase of the Cosmos' coming into being? If the Cosmos was created the way recent theory supposes it to have happened, it must in any event have been very fundamentally different from any physical processes about which the scientists have direct evidence. We must be permitted to wonder whether the whole intellectual exercise of extrapolation backwards is successful or not. Maybe the Big Bang was not quite what the current generation of astrophysicists believe that it was. But even if it was so, why should the laws of physics be the same before there were any stars or even any atoms?

It is trivial knowledge that many specific laws of both physics and chemistry apply very differently at different levels of ambient temperature, for instance, and specifically when the various elements are solid, fluid, or gaseous, as all elements can be at some temperature or other. No one expects solid-state physics to be the same as gaseous physics. The temperatures at which elements are solid, fluid, or gaseous are also very different even between elements that are close to each other in the periodic system or occur in analogous positions in that same system. Many gases can also be chemically active or inactive at different temperatures, specific to each element at its gaseous stage.

These "phase transitions" in physics have their analogies elsewhere. Thus, for instance, in biology there are dormancy, stupor, reproductive processes, and metamorphosis (as in insects).

From such analogies it would follow that at densities vastly different from those we know empirically, laws such as those of the weak and the strong nuclear forces—or of gravity or of electromagnetism—could very well be different. For electromagnetism this is confirmed by recent findings on superconductivity at very low temperatures. It may very well remain problematic if the physicists can ever formulate what those laws of physics may have been at a stage in cosmic development that appears to have been entirely different from all that can be observed even indirectly.

The problem calls to mind some recent statements about "anthropic" principles of knowledge, as discussed in a previous essay. Where conditions are unlivable, humans cannot tread, and where they are totally hostile even to the elementary conditions of our present world, even the human mind may be unable to tread. We noted this in the first essay about "nature's traps" for our mind, as exemplified by the initial difficulty of discovering nuclear physics. The problem of the origin of the Cosmos is a neat analogy—with a vengeance.

At the root of the search for absolutely universal laws of nature is the problem of explanation versus description, already alluded to. We shall, of course, return to this, and more than once. Meanwhile, let us seek enlightenment from contemplating some specific laws of nature.

SPECIFIC LAWS

Besides a few general laws that we just discussed, there are also innumerable specific laws of nature governing one thing or another in the Cosmos, from the movements of atoms and subatomic particles, to the chemical rules that command the build-up of compounds both inorganic and organic, such as proteins and genetic material and their effects on heredity, for instance. On reflection, it turns out that every discreet phenomenon that we can identify and give a name has some regular necessities of their own. Those that cannot be derived from the general laws, we can classify as specific laws of nature.

Science consists, to a large extent, of a search for invariances. Invariant phenomena or processes are supposed to reflect inherent necessities. If they did not—if the phenomenon under review were subject to unaccountable variation—then science may not be able to handle the occurrence. We found plausible instances among subjective experiences in aesthetics and religion, possibly also in parapsychology. When something appears to occur in random variation, then this can only be described statistically, but the occurrence of the minority case—the improbable case—is and remains indeterminate.

If an invariant phenomenon is defined with sufficient clarity, then the invariance can be said to reflect a law of nature, whether specific for this class of phenomena or more generally applicable also to other classes of phenomena. Among the plethora of specific laws, we usually designate some as belonging to physics and/or chemistry, with the latter shading over into biology. Laws of physics and chemistry act as constraints on living matter. They seldom define all that goes on within living organisms. Much less can it be taken for certain that chemistry would explain all that occurs on the mental plane, culminating in the human mind. On that plane, individually different variations become so frequent that laws of nature cannot always be invoked as complete explanations. Attempts at doing so usually end up in some kind of logical circle that purposely disregards the unknown. We now know that there can be events without compelling cause because of random variation in causating factors. More of that when we come to the limits of science in history.

LAWS OF PHYSICS AND CHEMISTRY

Already at the level of chemical elements, we encounter invariances that cannot in all respects be derived from the known properties of the constituent particles. It is true that the periodic system of elements allowed Mendeleyev to predict some properties of chemical elements not yet discovered, and Mendeleyev predicted that they would be discovered. His predictions came true; the predicted elements were eventually found, and they have chemical properties that include those he had predicted. But that is not all there is to chemical elements. Elements have many properties that cannot be predicted or concluded from their places in the periodic system, or by their composition of protons and electrons.

This is especially true at the level of chemistry, that is, when elements combine into compounds. If it had been possible to conclude all the properties of compounds from the places of the constituent elements in the periodic system, then we would not have needed the vast laboratory resources by which the specifics of chemical compounds, and their ways of reacting with each other, are now painstakingly explored, often with approximations later to be refined in still more advanced experiments.

All known determinants of the properties of physical particles and energy units, and of chemical elements and their compounds, have their definable strength and their empirical limits. Among the elements, for instance, odd-numbered and even-numbered periodic numbers lead to different kinds of properties, depending on how the "electron shell" is open or closed toward the configurations of other elements. It is certainly not by chance that all the chemically inert gases (the noble gases) are even-numbered, at 2 (helium), 10 (neon), 18 (argon), 36 (krypton), 54 (xenon), and 86 (radon)—there may be exceptions at some temperatures. Being chemically inert, their properties are more readily specified than is true of elements that can enter into compounds. Most elements can, but some only in a limited number of compounds as in the case of gold, the noble metal. Some others, notably hydrogen, carbon, nitrogen, and oxygen (all of them light elements) can be parts of vast numbers of compounds, many of them extremely complex. Carbon exceeds all other elements in the ability to enter into compounds. In general, heavy elements are less prone to form compounds than are the light ones, but this is true only in broad features, not in every detail, as the case of the noble gases would have shown.

Puzzling to many of us is the lantanid group, fourteen elements with periodic numbers from 58 through 71—also known as the rare-earth metals. The puzzle is that among these elements, there are only slight chemical differences. There seems to be no particular explanation for this curious occurrence. Any attempt at explanation risks very much to represent some kind of logical circle.

There are also elements that are unstable, prone to decay rapidly into some other element, and thus hardly occur in nature. This includes technetium (43) and astatine (85), as well as the "transuranians" (those with periodic numbers higher than 92), beginning with neptunium (93), plutonium (94), and several still heavier elements known only from laboratories.

And so it goes with much of the chemical variations. The properties of compounds can to some extent be anticipated from those of the constituent elements, but many of theses specific properties of compounds just have to be explored and described. The invariances that are empirically found in them will, in turn, serve as explanations for what goes on in even more complex compounds. Some simple compounds can behave in ways that resemble those of elements, thus ammonia (NH_3) and the simple cyanide mole (CN), the base material for some very potent poisons.

The ways in which molecule designs can affect the chemical properties of complex molecules is highlighted when both polymers (multiples of the same design) and isomers (different designs engaging the same combinations of constituent atoms) can have properties that are different from those of the compounds of which they are polymers or isomers, respectively.

Trying to gain some order into all these effects of combining designs into still more complex ones we will, time and again, run into what appear to be organizing principles. It will be important to understand whether such apparent principles represent, in effect, invariant general laws of nature in themselves, or maybe only represent yet another set of limited-range invariances, to be explored and cataloged in analogy with the catalog of physical and chemical properties of elements. Let us discuss, in turn, some cases of leverage, catalysis, and synergy.

LEVERAGE

We chose leverage to represent the "simple machines," which also include the block, the sliding plane, and the wheel. In leverage there is complete explanation (or so it seems) as to the gain in effect, by the arms of the lever having different lengths on either side of the pivotal point. What is lost in distance is gained in power or effect; the equation is completely consistent, and no unknown component enters the phenomenon or its explanation.

The only limit is that of the materials involved: They are assumed to be hard and unyielding. That may not always be so. Leverage on a large snowball is not quite the same as that on a boulder, even of the same weight. The equation may have to be modified to allow for viscosity as well as specific weight and other circumstances that may affect the application of leverage. Similarly, the anecdotal (and fictitious) case of a little girl preventing a railroad train from moving with her bare hands—because she holds a lever that is suitably proportioned—would also have to make allowances for the properties of the little girl's hands and arms. Not that it would make the case impossible, but the equation would not be exactly the same as with that of a steel robot holding the lever.

Leverage has its analogies in the other simple machines, as well as in the array of mechanical rules applying in the physics of gases, liquids, and solid substances, for instance. Those for gases are the simplest because gases tend to fill the three-dimensional space evenly, so that gas pressure extends equally in all directions and is proportional to mass and density. Liquid-state and solid-state physics are more complicated, but on the whole, they can be explained in terms of mechanical principles and their mathematical relations. If there are any as yet unexplained things among these various kinds of physical relations, these may or may not be explained eventually. In this area, we have no strong reason to believe that some of the explanations might be strictly ad hoc, for one case or another, without reference to general mechanical principles.

Looking at catalysis and synergy, we will have to become accustomed to less generic explanations.

CATALYSIS

This term covers a range of phenomena wherein a chemical reaction is speeded up by the immediate presence of some chemical substance (an element, an alloy, or a compound) that is not in itself a part of either the starting material or the end product of the catalytic reaction. A substance serving in this way is called a catalyst. Catalysis has wide applications in organic chemistry—in the lives of most, if not all, living organisms. It also has vast applications in both experimental and industrial chemistry. Some catalytic phenomena have been known since time immemorial—the sourdough example in the Bible refers to a common-sense experience, to symbolize where small causes could set off large effects.

Catalysis includes some features of general law and some of specific properties both in the catalysts and in the materials being transformed by catalysis. The latter we can classify as very specific laws of nature, laws that in precisely that form only apply to a limited range of analogous phenomena.

A general theory of catalysis includes the finding that the catalyst at first enters the chemical reaction, which it sets off and thereafter leaves the new compound it has helped to bring into existence—the catalyst is reconstituted and can serve in new rounds of catalysis. Which materials are catalysts and for what chemical reactions appears to obey rules that are specific to each catalyst and maybe also to each reaction that a given catalyst can help along. Such facts are invariant one by one. They cannot be completely or systematically derived from the known properties of the materials involved. If they could be so derived by general principles, huge amounts of laboratory work now being routinely performed would be unnecessary. This lack of general theory also includes the "efficiency" of catalysts, which is twofold: speed and durability. Some catalysts bring about the same reaction faster than others. Some are reconstituted completely, or nearly so, after each round of catalysis, and thus have a long (maybe even unlimited) lifespan as catalysts. Others are less than completely reconstituted and thus have a shorter life span as catalysts—they must be replaced or supplemented after some time of use. In industrial chemistry, much applied research is spent in search of the best catalytic material for a given chemical process, having regard to both speed and life span. In other cases, the search is for the least catalytic materials—that is, when the catalytic reaction is not wanted in the process or mechanism being developed. In biology, auto-catalysis is among the important feedback mechanisms that regulate the growth, the development, and the stability of living organisms.

In some contrast to the mechanical rules of physics, the chemistry of catalysis thus includes both some general principles and a vast array of specific properties of various materials. Once discovered, a given catalyst allows the chemists to explore and describe its invariances, as they can with regard to numerous other properties of elements and compounds. But a large part of what the chemists have thus found out (or are about to find out) consists of highly specific laws of nature. Only a relatively minor portion of all this can be derived from general principles.

The extreme type of specific invariances are those included under the general term of synergy, or synergistic effect.

SYNERGY

Synergy is now widely recognized in many areas of knowledge, perhaps most often in organic chemistry and biology. It also has numerous applications in human affairs, in what is often termed social science.

The concept of synergy has been defined as the case where "behavior of whole systems cannot be predicted by the properties of the several parts of the system" (Fuller 1972; Edmondson 1987). The same thing has recently been labelled "emergent property" (Calvin 1990), which is said to be a property being more (or less) than the sum of its parts. In other words: The first axiom of logic, which states that the whole is equal to the sum of its parts, is valid only in a strictly quantitative sense. It is not always valid of qualitative properties that can differ from the sum of parts in ways that cannot be predicted or stated a priori, but must be explored separately for each combination of parts. Once explored, a given synergy is expected to be repeatable and predictable as an invariance in its own right. The logic of qualia (qualitative distinctions) often escapes the purely quantitative logic of mathematics.

The choice of term may be debated. Synergy seems to indicate that the qualitative innovations were in some sense predicated on a synthesis of energies. This may reflect the often-used assumptions that energies may be qualitatively different, as they appear to be functionally. From our discussion of matter and energy we have the conclusion that energies as such can be entirely featureless, and that specific energy forms gain their nature, and the kind of effect they can set off, not from quanta as such but from the design in which the quanta are arrayed and move. Thus, synergy should really be understood as a synthesis of designs—each of them may be qualitative to begin with, and then it ought not to surprise us that a larger design that includes two or more designs of a lesser order could also emerge as something qualitatively new. "Syndesign" might be a more accurate term, but at this stage it appears risky to introduce a new term—there is enough confusion about this sort of thing to begin with, so we will stick with the inherited term, logically inadequate as it may be. "Emergent property," on the other hand, appears too vague to really enlighten those who need it.

In explaining emergent property, a basic instance being quoted is that of color perception, first described by Thomas Young in 1802. A combination of two colors sometimes leads to a perception that does not strike us as intermediate between the two colors being combined, but as something qualitatively new. A more recent example labelled emergent property is in the emotional character of a melody (Calvin 1990). A sequence of physical properties in musical tones meeting a sequence of chemical processes in our brains and a sequence of cultural traits in our minds—all of this gives rise to a new experience that cannot be reduced to its components.

These examples of emergent property both refer to subjective experiences in the human mind. Not that we want to underplay this class of phenomena—far from it, as our subsequent discussion will testify. But it still is of some consequence for our thinking that the more generally accepted concept of synergy applies also to phenomena and processes that can be observed in strictly objective ways, without, for that sake, rendering the synergistic phenomenon any less of an innovation. The vast application to both chemistry and physiology allows us to surmise that synergistic change must be even more important in human physiology and culture.

Looking down from the welter of applications in organic chemistry, we can also conclude that the same principle is useful in explaning (that is, really, cataloguing) how the chemical elements and their more elementary (often inorganic) compounds gained their specific properties—those properties that could not be completely concluded from the nuclear configurations of atoms.

Even a complete computer model only describes how a synergy is built up—it does not explain the qualitatively new outcome. Now it becomes evident that explanation equals description of invariances. Each synergistic invariance can be catalogued as a specific law of nature. In logic, an established and well-known synergy is treated like a "black box"— we know what goes in and what comes out, but the link between the two is only described, not explained in any other sense. It is as unexplained as the existence of the Cosmos itself.

We may now turn around and resume our discussion of the general laws of nature.

THE "GENERAL FORCES" AND THEIR LAWS

Of the generally debated four forces, only one has been explained or elucidated in any satisfactory way: The electromagnetic forces (in the wide sense) have all been shown to consist of photons traveling at or near the speed of light, some with spin (as the electrons), all with different wave patterns or wave functions. It is these various spins and wave patterns that make the differences between electricity, x-rays, heat, light, and so on. The spins and wave patterns are the designs of these several energy forms, ruling over the elementary building blocks, the photons. Without any wave pattern or other superimposed design, the photons might not even be real or they would in any event not be likely to have any strength, notwithstanding their movement at the speed of light, if that is the case—if such designless photons exist in some sense. Again the neutrino comes to mind, but then the physicists seem to have very little positive to say about neutrinos.

Recently, the rapidly expanding field of laser technology has given us an impressive object lesson about the immense consequences that may follow when a new kind of design is imposed upon elementary particles. A laser beam differs from ordinary light by its design, not by the quanta of which both are composed.

The other three general forces are less well explained. Let us begin with the "strong nuclear force" that is supposed to be what holds together the large subatomic particles—the protons and the neutrons and their counterparts in the world of antimatter, such as the antiprotons, and so on. What this force should explain is the exceptionally strong cohesion of the protons. The neutrons as such are somewhat less stable— they often decay into protons while splitting off one electron and one neutrino each. For the protons, those who think of force as energy and energy as conveyed by specific particles have invented a hypothetical one dubbed "gluon," which should deliver the cohesive force of the extremely stable protons—their glue, apparently.

No experiments to date seem to have come across any gluons. Nor, in independent existence, have the "quarks," of which a few are supposed to add up to a proton or a neutron with specific effects as to electric charge, been discovered. If we read the literature correctly, this should be the important difference between a proton and a neutron, the latter being without any such charge. How then is a neutron rendered into a proton merely by shedding two minimal particles?

Assuming an otherwise unexplained force to be due to a hypothetical particle is perhaps logical at a certain stage of science. To an onlooking layman, such constructs get an uncanny resemblance to the "epicycles" by which the late Ptolemaians tried to explain away inconsistencies in their system. Rather than hypothesizing about an unobserved class of particles, would it not be much simpler to regard a proton as a contraption of high stability?

The world of mechanics includes innumerable examples of contraptions that are held together, not by any specific force or energy charge, but by their design. As an instance, think of a chain made by links that can be switched open or shut by a snap of the finger by anyone who knows how. Such a contraption does not need any specific quantity of dynamite to be blown open. Manipulating the design is all it takes, and that does not require anything near the energy equivalent of the strength with which the contraption resists destruction by brute force. There is indeed an empirical reason to propose a hypothesis that the proton is a very stable contraption: the manner in which a proton and an antiproton, upon contact, blow each other up, each disintegrating into 1836(?) photons. The only difference between a proton and an antiproton is that the proton spins leftward, the antiproton to the right.

The hypothesis of the proton as a contraption where 1836(?) photons are gripping onto each other, in a unity that is unbreakable as long the leftward (or rightward) spin remains undisturbed, is made easier if one assumes the photons to have the form of strings of some articulate shape. This shape may or may not be different in different wave patterns, when photons move away at the speed of light. As they keep gripping each other inside a proton, they may or may not have yet another shape than they have as free photons. No assumption of this kind is needed in order to ideate the proton as a contraption, rather than as a unit ridden with gluons to hold it together. No specific energy form would be required, only some specific interior design of the proton. So, maybe, the strong nuclear force is not at all any specific energy type, but rather a design feature leading to great stability and requiring great force to break it up.

The weak nuclear force, which is supposed to hold together the atoms and their nuclei, may also not be any specific energy form. It could merely be another design in which the subparticles hold each other. Even though some energy is released in nuclear fusion, it still might not be necessary to assume any specific quanta, or any particular kind of particle, to hold the nuclei together. Cohesiveness of design could answer the same requirements. Blowing up atoms is another affair—that may require force.

Nor do we have to count the spin of the proton as a force, since this is merely an equilibrium feature that consumes no energy and thus is able to go on indefinitely without being any kind of perpetuum mobile. The absurdity of a perpetuum mobile, in earlier discussion, was that such an apparatus was assumed to be doing work indefinitely. Doing work is changing something else by the action of an agent-at-work, the latter being worn down or used up through this activity. The ongoing spin of a proton, or of the electrons around the nucleus, does not do any work. Such movements merely maintain a system unchanged, which is the opposite of doing work. Similarly, the rotation of the Earth and the other planets around the Sun does no work. These orbital movements merely keep a system intact. Such movements have some consequences, of course, such as the way solar radiation is received in seasonal variation. The movements of the Moon also relate to tidal movements on Earth, and so on. But the orbiting movements of planets and moons are not work—they are merely resisting change. The same must be assumed to hold for the protons spinning forever leftward and never seeming to go to pieces, except from meeting an antiproton or some other process inside the complex artifacts through which modern physics does much of its research.

Where does all this leave the fourth general force, gravity? Nowhere in particular, that is true; but then many physicists appear to be as stumped by gravity as ever (Chaisson 1988). Newton and Laplace did not attempt to explain gravity; they merely described it. The remaining difficulty has to do with the apparent "force at a distance"—how could such a thing be possible? To bring this apparent mystery within the domain of received physics theory, another class of particles has been assumed, dubbed "gravitons," which in some way should be the harbingers of gravity attraction. This is just another hypothesis, nowhere yet confirmed; so maybe this is yet another modern day epicycle. Einstein's "curved space" is also not always helpful—certainly not for someone who happens to be in the way of a ton of bricks falling down because a steel wire in a crane happened to snap just then—really, Newton's apple writ large. Gravity is often referred to as a weak force, but that depends on one's viewpoint, as this example would demonstrate. The problem of effect at a distance is what the assumptions of gravitons should rationalize, and these particles never having been detected is blamed on their being few and far between. The falling ton of bricks might then obey the command of rather few gravitons, and the falling apple even fewer. At the other end, Einstein's curved space is appealing in regard to celestial bodies and their staying in orbit as the limit value between the escape velocity on one hand, and the inferior velocity at which the orbiting alternative is broken and the visiting body (comet, asteroid, meteor, whatnot) is forced to fall into the superior body—whether star, planet, or moon.

Recently, the movement of our galactic system, at high speed, is attributed to gravity pull from some distant "Great Attractor," not otherwise known than from this intergalactic effect. Indeed a weird application, whether of curved space or of gravitons.

With all this, we must question whether the four forces really are so many discrete forces. The nuclear forces, we have argued, may well prove to disaggregate into a number of separate phenomena of design rather than force. Each of them may be governed by some specific, yet rather general, law of nature.

The best established and best explained of the forces of nature is electromagnetism, but this seems to be one force only inasmuch as moving photons are the common building block. But the many variants of wave function and spin seem to dissolve electromagnetism into a number of discrete processes, each defined by the specific laws of nature governing their designs of wave function and spin. All the light particles "on the loose" have in common their centrifugal movement, away from a source of origin, but nothing in particular pushes them, no extra particle is thought of as causing this centrifugal movement. The traveling photons themselves *are* the force that makes them move. Instead of some ultimate cause, all we find is a descriptive account of what goes on.

Turning around to look at gravity, we find the reverse. This concerns bodies heavier than those of radiation, and their movement is centripetal. Assuming them to be pushed by a specific kind of particle, the graviton, would break the analogy with electromagnetism in which the moving particles are their own source of movement. We thus propose to revert to Newton's stance, regarding gravity just as a fact of existence, to be described but not to be given any ultimate explanation other than its very obvious existence.

The onlooking layman must also be forgiven for suggesting that maybe—just maybe—Einstein's equations are not the absolute last word; they may be an extremely close approximation that, however, may not be sufficient to support all the bold conclusions that present-day cosmologists are drawing by extrapolating the consequences of Einstein's equations to the hilt (Lightman 1991).

In reductionist thinking, it is assumed that all the specific laws of nature could, in principle, be explained as complex applications of the elementary forces of nature. High complexity is all that prevents us from solving all these advanced equations—this is the gist of a statement frequently attributed to Paul Dirac. This is, of course, no more than an arbitrary assumption. There is no proof to support this assumption.

A mere reference to Occam's razor will not suffice. In the real world of science, many of the specific laws of nature must be treated as such—they are just there. They are no more "explained" than the ultimate "single force" would be, if such could be demonstrated.

VIRTUAL PARTICLES

We have mentioned some hypotheses on particles that have not been found experimentally, but are rationalized as "virtual" particles—they do not really exist, but things work as if they existed. Virtual existence has been mentioned already in the case of certain images on a camera lens. The concept is also known from electromagnetism: "Virtual photons" in some cases close the circle of such phenomena. Now the concept has also found application in something called quantum electrodynamics (QED) to cover those hypothesized gluons and gravitons and still other particles not yet found but considered necessary to explain what is going on.

Most of the popular literature on advanced physics seems to sidestep this type of problem, just saying that the gluons and the gravitons are hard to detect. In a central and highly authoritative treatment of QED, the category of virtual particles is defended on the ground that the theory works through mathematics (Feynman 1985; Livingston 1996, pp. 265–68).

Founding an advanced theory on mathematics alone, without any direct empirical observation of the basic concept, seems to this lay reader as merely a way of closing a logical circle. Our purpose here is not to propose an alternative theory of quantum electrodynamics (perish the thought!), but rather to set the record straight: The physicists have stumbled across something they cannot explain in materialist terms, and so just wish the problem away by using mathematical solutions. Some of these are, quite rightly, called "strange" or even "weird."

Would it not be safer to admit ignorance? Would that not be needed to clear the decks for a new approach to some phenomena that seem to defy current analytical techniques?

A WEB OF NECESSITIES

All the many laws of nature, from the most general to the most specific ones, together form a network within which reality is able to unfold. This extended cosmic design keeps the Cosmos existing and functioning the way it does. It does not help us foretell all that will or can happen. Randomness and low probability in chaos offer loopholes for unpredictable events.

We have no reason to regard the existing and observable laws of nature as being in any sense arbitrary. Saying that in a different cosmos the laws may be different does nothing to weaken the authority of the laws as they exist and function in our cosmos. Whatever does not conform to these laws is not *viable*, could not come into existence, or, if occasionally brought about by random events, would soon perish—like the building drawn by an incompetent architect. Such restrictions apply to some chemical elements, such as technetium and astatine and some extreme transuranians. It applies also to a multitude of theoretically conceivable compounds that cannot form—there can be no helium oxide in any physical circumstance known to us. The same applies to hypothetical life forms that could not go on living beyond a brief moment of paradoxical existence in some marginal circumstance.

Viable designs consist of applications of laws of nature. These are given, not arbitrary. In artifacts (such as buildings, etc.) the design must exist prior to the specimen in which the design is embodied—that much we could establish in the essay about design. We are now forced to conclude that in our cosmos, the way it functions and as far as our knowledge goes, the laws governing viable designs must also be thought of as existing prior to the coming into existence of any particular specimen or class of embodied phenomena (Davies 1988).

We need a caveat for "known circumstances," especially for the more specific laws of nature such as those of biochemistry. We had an object lesson not long ago when it was found that, near mini-volcanoes on the ocean floor, certain previously unknown organisms can live and thrive in water where the temperature exceeds that where water usually boils (100°C., 212°F.). The explanation is that, at these great ocean depths, the pressure from all the overlayering water prevents formation of steam (Cone 1991). What usually renders boiling deadly to all living organisms is precisely the formation of steam that must of necessity rupture all organic tissue. High temperature as such can apparently be tolerated as long as there is no actual boiling.

But this case does not render any of the known laws of nature invalid. It just warns us that, under some ambient conditions that we have not experienced, there may be variant laws not yet available for our observation.

The object lesson should, however, be heeded also in astrophysical cosmology. The warning of Hans Bethe comes to mind: We should not be sure that we can anticipate all that will happen or all that will be discovered in ambients that are in important respects different from anything about which we have well-grounded empirical knowledge.

VALID EVERYWHERE AND EVERYWHEN?

We are led to reflect on the universal validity of the many laws of nature, even given the *ceteris paribus* caveat referred to above. Does it still have any meaning to suggest that the known laws of nature were not valid in the earliest, largely (or, rather, entirely) unknown moments of the Cosmos—assuming there was a beginning, by Big Bang or otherwise? That depends on how we handle the ceteris paribus qualification. Obviously, the ambience was then rather different from anything we know. For that reason we should not expect the laws to be, or to function, as they are and function in the Cosmos we know from empirical evidence.

We may grant that there were many ambient circumstances quite different from those we know, so that many detailed laws of nature may have come into play differently in some analogy to what happened to those hot-water creatures on the ocean floor. But can we, for that reason, assume that Big Bang (or whoever or whatever) could at the moment of creation have *chosen* laws other than those that actually came about? Might the laws, starting from Moment One, have become quite different from those we know, even ceteris paribus? Did the initial creative event (of whatever description) have anything like free will as to the laws under which the Cosmos was to function?

There we have to stop. If we assume the laws of nature to be due to some arbitrary decision at the initial moment, such an assumption would certainly break all that our intellect is able to take. To any duration of cosmic time when the laws of physics and chemistry were as we know them, with whatever ceteris paribus qualifications because of variant ambients, we must perforce assume that the laws steering the physical and chemical processes antedated their specific applications. Just as with artifacts, the laws that rule over material phenomena and processes did (and do) in each case exist (in some sense) before they actually apply.

The laws of nature are, in principle, a domain independent of time, space, and quantum. Together with the inventory of viable designs of things they form a treasury of tools—the tool box of creation. Creating anything viable will depend on drawing tools from this treasury.

LAWS OF NATURE OR OF LOGIC?

It remains to reflect on the terms by which laws of nature are explored, understood, and explained. How problematic this can be will be evident when we learn that mathematicians cannot agree as to whether the statements of mathematics reflect what exists "out there," or are merely approximating constructs of human minds (Kaufmann and Smarr 1993).

Even the latter interpretation of mathematics does not render our reading of the laws of nature arbitrary or void of objective realism. In some ways our understanding of nature does reflect nature "out there," else we could not achieve the degree of mastery over nature that we have achieved and are achieving.

This may not be the ultimate truth, however. What mastery over nature proves is a certain degree of approximation sufficient for engineering results. Euclid's geometry has been very useful for many practical pursuits even though it does not obey the strictures of non-Euclidean geometry. But engineering results require only that—close approximation. This became even more evident from recent results on "fuzzy logic" (Zadeh and Kaprczyk 1992; Kosko 1993). (The fuzziness here refers to reality being fuzzy—there is nothing fuzzy about the logic.) Engineering does not require absolute truth in the sense of there being no further refinement possible.

As regards mathematics, a lay observer may be excused for suggesting that the choice of answer—mere mental construct versus true reflection of what is actually "out there"—may not be absolute. Let us consider the case of digitalization of curved motion versus the case of geometric proportions.

Digitalization of curved motion brings to mind the classical case of the Eleatic school of Greek thinkers, which declared that motion was impossible because it would have to consist of infinitely numerous subunits, and this would take forever to come about. The logical fallacy of the Eleatics has been exposed many times. The trajectory of a flying arrow, or of an artillery shell, does not consist of discrete subunits—it is a single phenomenon to be understood as a whole or not at all. Digitalizing it is approximation, not complete or absolute truth. It answers practical concerns, not theoretical perfection. There are really no definable moments or loci on the curve because the curve is a single phenomenon. We may even suggest a modification on Heisenberg's famous uncertainty statement. Instead of saying, as he did, that we cannot at the same time establish both the locus and the momentum of a particle, we may elect to say that a moving particle can only be established as to its momentum (time taken to get from here to there), but never exactly as to its locus. Being in continuous motion, the particle never occupies any definable place for any definable moment of time. Continuous motion is a "whole" phenomenon, not to be reduced into definable parts.

Geometric proportions offer some quite different cases. The proportion between the circumference and the area of a circle, as mediated by the π (pi) constant, must be accepted as existing out there no matter how incomplete our numerical calculations may remain. The analogous formulas for proportions within ellipses, parabolas, and hyperbolas must also be accepted in a way that mathematicians can do nothing to change.

Likewise the trigonometric functions: Granted the variants dictated by non-Euclidean geometry, the sine and the tangent must be accepted as existing out there. As proportions, they are not dependent upon any particular system of numerical representation.

By analogy, our formulations of the laws of nature may fall in one or the other category—some may be mere constructs of human minds while others really reflect what goes on out there. The law of gravity is obviously there, no matter what. A tree in the forest that falls does so independently of whether anyone observes it or not. The properties of helium and of carbon, for instance, are as they are; no amount of reformulation can render helium other than chemically inert, nor make carbon other than capable of immense amounts of chemical compounding. Adding the ceteris paribus caveat does not change the basic situation.

At the other end, many of the received laws of electromagnetism and of nuclear physics are certain only as constructs of science, capable of giving people a certain degree of power over nature. How definitive these constructs may be is not known with any confidence, nor is that likely to become known in any foreseeable future.

REDUCTION TO THE ABSURD

An important feature of modern science is the drive toward reduction of complex phenomena into their component parts, as a way of explaining the complex from the less complex. The ultimate in reductionism is the search for a single ultimate force of nature, a force that is supposed to have existed "in the beginning" (just after the Big Bang), later to be differentiated into the four basic forces.

We have already argued that synergisms cannot be reduced to components. They can be so reduced only in a mechanical sense—yes, they have components all right, but reducing the synergy to its components merely destroys the synergy. The qualitative outcome of a synergism is never explained—it cannot be reduced into subunits of qualities that constitute parts of its quality. For instance, saying that the human soul is explained by the physiology of the brain (Crick 1994) is about as convincing as it would be to say that the emotional quality of a cathedral vault is explained by the bricks and the mortar of which it is built.

This stop sign in the path of reductionism should not astound as much as it sometimes appears to. Even the most extreme reduction has to stop someplace. Suppose cosmological astrophysics was to succeed in reducing the four forces of nature into a single original force. So what? That ultimate force itself could never be explained. Explaining requires some ground to stand on, and there would be none beyond that fateful first moment of existence. A topnotch scientist, musing over this ultimate failure of explanation was heard to add, in what sounded like a tired voice, that we have no explanation as to "why it all bothers to exist." The fact of there being a cosmos, instead of nothing at all, evidently has no explanation.

We may now turn around and propose that the same applies to the entire complex web of all the laws of nature. The innumerable specific ones, finding expression in one invariant synergism or another, all just "are there." We cannot ideate any explanation of it all, or even of a single one among the synergisms of organic chemistry, for instance. The hectic chase of reductionism breaks down and makes room for the calm acceptance of descriptive observation.

BEYOND SCIENCE

This essay has been about the relations governing processes in our observable world of space, time, and quantum. Even this general definition—which limits as much as it clarifies—has to be tempered by the insight that many of the findings of science are approximations. The whole modern drive toward mastery over nature has somehow placed the power over nature above abstract truth, and so we should maintain a distinction between engineering and science. The often repeated statement that a scientific finding proves itself because "it works" is really the criterion of engineering. Recent departures into the vast domain of fuzzy logic underscore this distinction of prowess versus truth. The statement that probability mathematics is merely a subset under the larger concept of fuzzy logic underscores the point still further (Kosko 1993).

Even with these kinds of caveats, there are vast domains of possible reality that neither science nor technology can reach, as we discussed in one of the previous essays. We now want to pull together our conclusions about designs and processes in order to see what kind of world these represent. Enter the domain of Greek *logos*, of predefined patterns of existence.

SEVEN

Logos: The Ways Things Must Be

Alles Vergängliche ist nur ein Gleichnis
(All that is perishable is just a likeness)

Goethe, *Faust* (Chorus Mysticus)

The notion that our reality is but a temporary reflection or incarnation of ideas that are eternal was already full-blown in the philosophy of Plato. Ancient Greek thinking gave us the term *logos* for the inherent, timeless design that decides which things are which in this world of temporary phenomena. The concept has been carried forward many times by the author of the fourth Gospel in the New Testament (St. John), among others. His use of logos has been obscured in most translations by the literal rendering as "word." In fact, his logos is the Platonian concept of ideas that antedate any appearance "in the flesh" and also survive it. The totality of all the logos in the world, so the gospel writer tells us, is God.

What follows in this essay is not an attempt to explore the ultimate metaphysical consequences of the notion that the designs of viable phenomena are in some sense eternal. That may be out of reach, anyway. Rather, we want to trace the significance of this notion for our ideas about knowledge and ignorance—the latter no less than the former.

Let us begin by summarizing some conclusions from previous essays.

DESIGN AND PROCESS

The design essay showed all that is real to have some features, some design. This goes for both objects and processes. Every "thing" in existence must exist in some specific way—must have some design—else it could not be real. Every going-on, every process, must also have some features, some "how" of the process, else there would not be anything going on. No process is possible without some specific modus operandi.

Standard materialism is based, ultimately, on underestimation of the power of design. Matter has commanded more elementary respect than spirit because matter are things where the design is relatively very stable. Stable objects can easily be mistaken as being more essential to the world's realities than unstable ones. On such logic, granite would be more real than radium; windstorms and fires more impressive than flashes of genius. On close reflection, we find the opposites to be true. Radioactivity can destroy or reshape the most solid minerals, and genius can find ways of harnessing both wind and fire. But the greater display needed for change tends to make the stable things, and the regular natural processes, appear more essential, more real.

To the contrary, the designs that overcome the most resistance—the most complex ones—are in this sense more "real" than the simpler, more elementary ones—in the sense, namely, that design makes reality. Hence, things with more complex designs are also more real.

The same goes for processes: Those with the most elaborate features in their modus operandi also are more real and can contribute more to the coming-into-existence of more reality. Organic chemistry is more powerful than elementary inorganic chemistry, for organic chemistry contributes more to increased design intensity of things—that is, to more reality.

Designs and processes in the real world (as distinct from "virtual" reality, which is discussed more below) are however constrained by a criterion of viability: the toolbox of creation.

TOOLBOX OF CREATION

We have already introduced this concept, which in itself is rather self-evident. It is self-evident that a great many arbitrary designs could never materialize into anything of more than fleeting consequence. The house drawn by an incompetent architect will never stand—not on our planet, not on any other planet either. What may happen to it in virtual reality we can discuss later. When all the incompetent designs are weeded out, the remainder of what could be ideated constitutes an inventory of what is possible or feasible in one context or another.

We must question how far the tools of creation are the same as the concepts used by scientists and engineers in their formulations and solutions of problems. We already alluded to this when referring to the debate among mathematicians as to what is out there and how it may relate to their formulas. We shall have reason to return to this distinction later. For now, we must repeat that a scientific formula that works is not necessarily identical with anything in creation's tool box—not exactly, that is. Science, to a large extent, proceeds by approximations, and these are not always exactly identical with the tools of creation. The difference is not always "operational" or of consequence for engineering. When discussing theories of knowledge and ignorance, we must retain the distinction between the absolutely true and the approximately true.

The widest yawning gulf between knowledge and reality is found in the problem of the human mind. Soul and consciousness seem forever to elude any strictly scientific approach, and yet they are always with us.

MIND AND THE BRAIN

Looking for the mind, search the brain. One of the classical materialists once wrote that the brain generates thoughts and emotions in much the same way as the kidneys secrete urine. More is now known about kidneys, and they turned out to be much more complex and refined than anyone imagined when that line was first written. The simile of a physiological organ function has been repeated recently, using the stomach as a parallel of the brain (Searle 1988, as quoted by Baker 1992). The soul is still sometimes represented as just a function of neurons (Crick 1994).

The relation between the brain and our minds or souls is, in fact, still unknown. Much research has been done on brain surgery, with great results for therapy, but we are still far from having any theory of brain function versus mind (Calvin and Ojemann 1994).

How the brain, that record-dense computer containing billions of gray cells and performing unspeakable numbers of operations per second, can generate thoughts and consciousness is still a complete mystery (Alkon 1992), presenting no fathomable analogy to the chemical filtering of kidney membranes and stomach linings. In the kidneys and the stomach, certain materials go in and an array of somewhat transformed materials come out. In the brain, some measurable impulses (and some nonmeasurable ones, too) go in, and out come our thoughts and emotions, leaving no indication as to what was transformed or how this was done.

The physiology of kidneys and stomachs is highly standardized and their outputs are closely regulated. Any major deviation from the norm spells disease. By contrast, the output of the brain is variable ad infinitum even within perfect health or normalcy. The same brain may think of poetry or pancakes, of science or sweatshirts, of cosmology or cosmetology. Choose your own examples. The content of what the brain handles is in no way analogous to the stuff that the body's physiological organs handle.

Materialist science, in its extreme form, took the easy way out by denying what it cannot tackle. What we know about subjective knowledge and incompatible knowledge should teach us otherwise. The simile of the lighted room ought to tell us something of the "how" of a possible new orientation.

True-to-form behaviorists deny the existence of consciousness because they see no instrumentality that could photograph it. Likewise, Soviet dialectical materialism used to declare the human soul to be just "matter in motion," as if such an expression could deceive anyone out of the anguish of death. In either formulation, classical materialism accomplishes nothing more, really, than to reflect its own utter inability to comprehend human reality where it is at its most important.

SOUL AS OBJECT

To fall in line with prevailing materialist and reductionist thinking, we should be able to identify what kind of an object the human soul is. It is there, all right; how else could we say that someone "is as hard as nails." One mind opposing another mind may be as impenetrable as one solid object meeting another solid object. Solid objects are subject to the laws of solid-state physics. Laws governing minds may be just as compelling, only we do not know or perceive very clearly what they are.

From the standpoint of science, a main difficulty is that we have no experimental, nay, experiential, way of proving the human soul or what it may consist of. Untold amounts of ink have been spilt trying to explain what soul is and how it comes that we cannot conjure it away by any feat of dialectic. Not being an object to be touched in any manner but figuratively, the soul or mind has sometimes been declared to be an aspect of our single psychophysical existence (monism) or a parallel, something like a permanent shadow (psychophysiological parallelism).

Recent science has sought desperately to explain the conscious mind on the basis of physics or chemistry. It obviously exists, but not as a function of any specific particles. Searching for particles to become conscious is like searching for how bricks and mortar can add up to the inspiring quality of a cathedral vault. The bricks do not add up to any such thing and neither do atoms or molecules add up to a conscious mind. We have not even the beginnings of a theory on what the physics of the mind might be—not any more than we have any physics theory of a cathedral vault or of a piece of music. Synergisms just are not reducible.

In the cathedral, it is the architect's drawing that rules over the bricks (or whatever construction material was used). Likewise, we must insist that consciousness is a primary principle that rules over the physical and chemical particles that fill a body and its brain. Even the search for a threshold of chemical complexity, which makes consciousness possible in us humans—purported masters of the world (Dennett 1991)—is likely to be just another anthropocentric illusion, the lighted room writ large. All that can react to pain must be conscious in some degree, else the pain signals could not do their job, which is to warn. Hence, consciousness "in some degree" must be pervasive in the entire animal kingdom (at least).

Instead of rehashing the old arguments, let us dismiss the whole previous debate and start afresh from what we really know. Soul is reality, else we would all be fakes. In fact, soul is more real than any object, for all their realities are known to us through our mental apparatus—mind and emotions, the whole works are necessary for our ability to know.

A DIFFERENT MONISM

To resolve the seeming conundrum of the mighty soul arising from the dust of lowly matter, we must reverse the perspective of materialist monism. We recognize that dualism is not intellectually acceptable. We also recognize that parallelism tends to leave us in the void as to what is primary—matter or spirit. Materialism says that matter is primary, matter comes before spirit; spirit must then be explained in terms of material components and preconditions.

We now propose a revision of the basic perspective. It is as if current materialism were consistently looking at reality through the wrong end of a telescope. That kind of perspective renders the picture of spiritual reality puny—a thin fringe on a massively material cosmos of stars, planets, and cosmic debris. This reduction effect of the reversed telescope is more pronounced, the more complex—and, hence, the more real—is the immensity of spirit in the world, which we try to study under the microscoping perspective of materialism's intellectual instruments.

Turning the telescope around, looking at matter from the vantage point of spirit, the world of spirit ceases to be puny—it becomes grandiose, the more so the more the part we are looking at is a domain of high complexity, intensive reality. It is grandiose, too, for all the vastness of the stellar cosmos is matched by the human spirit's ability to grasp it.

This turning of the telescope follows logically from the entire construct presented in the essays on design and process. We found that all reality consists of design in one way or another; without any design there can be no reality. The timeless validity of nature's laws should force us to take creation's toolbox as the starting area. It is the starting area in reality, too, for the tools of creation rule over the things to be created, right from the start in elemental quanta of rays and particles. Logos was there before any of the things and processes built up at its command. If we follow this lead, recognizing spirit as primary in relation to matter, we shall find a great many of the classical problems falling in line. Soul and consciousness cease to be mysteries.

Soul is elusive because it is the agent that does all the observing. As "creation in being" it can become an architect in some ways. In a computer simile, the soul can be understood as the program inscribed in the software of the mechanism. Soul can be viewed as architect of its own mind and personality (Kornhaber 1988). It is not battered together by others acting as "construction workers" at our expense. It is built from within, like the genetic code builds a growing organism from within, or as an architect builds a building from within his conscious conception of the building-to-be.

ORDER FROM CHAOS

The immense role of complexity and variability in mental functions forces us to contemplate how these are at all possible in a world that was supposed to favor simplification, downhill movement away from complexity toward entropy.

Against the prevailing thought of a downhill trend toward ultimate "heat death," we have recently witnessed a renewed recognition of uphill movements, spontaneously in the direction of increasing complexity. Some of this is indicated in the early formulations of systems theory (von Bertalanffy 1968, 1981; Zadeh and Polak 1969). From systems theory the step is not far to recognizing synergisms, as well as nonequilibrium phase transitions and self-organization (Haken 1978), including morphogenesis in biology (the coming-into-being of organic forms).

Another early departure in the direction of order from chaos is in the findings on fractal geometry, as first described by Benoit Mandelbrot (Mandelbrot, cited in Hall 1992; McGuire 1991). The trait of "self-similarity" in fractal geometry seems to support the conclusion that the fractal pattern that comes about in a given chaotic situation is in some sense preordained by the precise traits in the originating chaos. Ice flowers on a cold window pane offer an everyday example.

There has been much theorizing recently about the origin of order. The same theme has now begun to break out of the mold of the usual materialist thinking. Complexity means life at the edge of chaos (Lewin

1992). Chaos is seen as creative because its results are not preordained. Minute differences in starting points may prove decisive for the outcome, and thus we see self-organization (Davies 1988). The same line of thought has led to the statement that matter has an innate tendency to self-organize and thus to create complexity (Coveney and Highfield 1995). These writers also pay much attention to "emergent properties" (meaning synergisms).

This already comes very close to vitalism, and must upset standard evolutionary theory, which assumes usable mutations to come about only at random for no reason any more compelling than what pushes the many nonusable mutations.

Along similar lines we meet the notion of the holographic universe (Talbot 1991), followed by the anatomy of consciousness (Rosenfield 1992), and all the way to love and its place in nature (Lear 1990).

MIND IN NATURE

The turning of the telescope allows us to interpret mind and consciousness in the human species as a case at the upper end of a scale, rather than as something new in principle. The question then is no longer what degree of organic complexity (in the brain) is needed to generate consciousness—uniquely in the human species or more vaguely also among the "higher" animals. Rather, the question becomes one of a lighted room: What degree of intense consciousness, as in our brain, will outshine the consciousness of lower organs? The simile of the "dark" sunspots comes to mind here. How much animal consciousness did we lose when we gained the specific human degree of brightness?

Recent discoveries in biology have taught us some very interesting lessons about the coming-into-being of increasingly complex organisms.

Contrary to previous beliefs, this was done not merely by the accumulation of successful mutations in the existing genome. It depended just as much on symbiosis, interaction, and interdependence of previously quite separate organisms. The organelles in our body cells have been proven, to a large extent, to be old bacteria species that have turned into parts of much more complex organisms. One forward thinker has gone so far as to characterize the human brain as "a colony of bacteria" (Margulis and Sagan 1986).

This formulation is rich in suggestions as to what else can be traced through the turned-around telescope. The primeval ocean, for maybe 2 billion years, was populated only by bacteria. It may have had some characteristics of a loosely organized cosmic brain. The primeval bacteria had no clear species limits, and they exchanged heredity material in protosexual contacts. This slowly built up the ability to thrive in their fluid universe, even as they themselves gradually became more capable of living in more circumscribing symbiosis. In the much larger (and already quite complex) eukaryotic cells (e.g., amoebas), which may have come about something like a billion years ago, the organelles already cooperate much more closely than the bacteria in the primeval ocean may have done. As organelles, they lost the ability to live "at large" in the ocean.

Any place in between will show us intermediate degrees of interdependency, whether symbiotic or otherwise. The "silent sea" has turned out to be a very noisy place for the beings that belong there. The fish, presumed mute because they seemed so to human ears, have been found to be constantly immersed in a bath of communicative signals traversing the waters. The wild of the forest—even the semi-domesticate of the "urban forest"—is reverberating with signals, as, for instance, when birds, squirrels, and so on make known their apprehension of an approaching predator. Who is there to say if the jungle and the ocean, both loosely organized biological ambients, may be "conscious" in a degree that could not function among humans in our concentrated "lighted rooms"?

Let us open up the paradox of the limiting effect of our highly equipped minds. Seen from our perspective, the human mind is strongly characterized by its limitations, no less than by its higher organization.

MIND AND SOCIETY

Human society not only shapes our minds, it also sets limits to them by the indoctrination of childhood and adolescent education, and by the culture that dominates our immediate environment and frowns upon deviations. Education and culture are the origins of many intellectual traps that come in addition to those set for us by nature. We can perceive only what we have been taught that we can legitimately perceive. In such ways, education and culture decide what doors of perception will be open and which rooms will be lighted.

Education and culture include as much of a mental starvation, in growing humans, as the physical starvation inflicted upon the larvae of bees and wasps destined to become workers rather than queens. The mental starvation inflicted upon us by education and culture renders us useful to society more than to ourselves. To ourselves also, indirectly: "What is not good for the swarm is not good for the bee" (Marcus Aurelius).

But when education and culture are too successful at molding us into disciplined workers, they may also choke off their own wellsprings of innovation. The failure of success, because nothing was learned from it (Kenneth Boulding) has many historical cases, the largest maybe in Imperial China (which indirectly set the agenda for modern China).

Hence, the oddball, the "fool," the genius, and the gifted criminal—all the escapees from society's worker molding—become so many challenges by calling into question the validity of society's assumptions, by forcing a widening of its horizon, even by probing the functionality of its very system.

Some of the means of disciplining the mind are found in language and in other institutions, including some early cases of virtual reality.

LANGUAGE

Much has been made of human language as a facilitator of our mental activities. Much less attention has been paid to its role of limiting these activities. Characteristic is a school of research that has insisted on the common roots of all human language, reflecting some kind of organic determinants that are thought to decide how analogous all languages are supposed to be "in depth." There is supposed to be a "deep structure" common to us all. (Regarding this theory of Noam Chomsky, see Pinker 1994.) It is not clear whether this "deep structure" is assumed to be liberating in its universality or maybe universally confining on the human intellect for the same reason.

In the practical world of languages, what counts are their manifold differences, not the hidden similarities they may have deep down. The ancient Greeks labelled those who could not speak Greek as "stammerers" (*barbaroi*). Slavs of late prehistoric time went even further, calling speakers of Germanic languages "mute" (*nemet*) and, by contrast, extolling themselves as those who possessed the gift of speech (*slava*)—the Slavic peoples were those who could speak! This even within the Indo-European family of languages where the deep similarities are not very well hidden. Anyone trying Chinese, or Maya, or other remote exotic idioms will have to acknowledge how far from each other prehistoric peoples may have been.

The limiting effect of language is but incompletely offset by the learning of "foreign" languages. Calling them foreign points to an implicit designation of one's own mother tongue as "the real language." It is usually the only complete one for each individual. Even that domain of mutual understanding is variously fractured by slang and technical lingo, some of it submerged as the silent language of "inner speech" (K. Dovring 1997). Even less understood is the barrier represented by concepts not allowed by the dominant culture: Thus the American abhorrence against propaganda dies even harder than the now defunct taboos on sex (K. Dovring 1959).

OTHER INSTITUTIONS

Language is an institution, or a set of institutions—that is, a set of rules and standards mutually agreed upon and mutually adhered to. Other institutions are, for instance, in the systems of law, custom, and mores. Some of these spill over into the much misunderstood category of "virtual reality."

Law is the set of rules that the public powers have pledged to enforce. The pledge may be in the form of legislation, judicial decisions, or administrative practice. When not enforced, a law may become imperfect or even obsolete.

Custom and mores are not enacted or enforced by public powers, but they are instituted and maintained by the general public, acting and reacting in concert. They can often be as compelling as law. The origin and the abolition of such rules may be more gradual than those of law, but for that sake they are not less conflict laden.

Law, customs, and mores are not only helpful in telling us what we can do or think, but they are also confining because of what they tell us not to do or even think. There are telling examples from the meeting of different legal systems. A Norwegian lawyer trying to explain (1920s) the land law of Norway to a committee of French lawyers was met by a brick wall of incomprehension: These things do not exist! They did not exist in Roman law (which was the basis for the French Code Civil), hence they could not exist at all in the minds of French lawyers. From Roman law we have the unbashedly absurd statement *Quod non est in actis non est in mundo* (Whatever is not in the documents of the case does not exist in the world). In such an extreme instance, positive law is revealed as a case of virtual reality, contrasting against the case of "natural" law, where all must obey the tool box of creation.

When the legal system parts company with the realities of the actual world, it may get lost in a world of virtual reality that still can function after a fashion. Even absurd rules may function because there is force behind them—force of police, force of public opinion, and, worst of all, the force that comes from the systematic outlawing of alternatives.

A clear-cut case of a social institution functioning as virtual reality is in the monetary system. The monetary unit does not really exist as such "in the flesh." Any and all embodiments of the dollar (or whatever form of currency) are arbitrary, maintained by accepted rules of law and custom. There is no object that in itself is "dollar." The various manifestations in coins, bank notes, bank drafts, and so on are all incidental to the system application of a fictitious unit. How fictitious it is becomes evident from the fact that Congress and the President can demonetize any coin, note, and so forth. By a few strokes of the pen in Washington, such-and-such piece of paper ceases to be money—in the hands of maybe millions of people.

And yet, the monetary system is among the most powerful tools of public policy at all levels. Shifts in currency futures can (and daily do) easily transform the money value of papers held on the other side of the globe.

VIRTUAL REALITY

The term virtual reality has come into wide use recently, and is usually used in reference to certain advanced computer techniques that seem to allow some people, with the help of rather complex equipment, to experience sensations and apparent events without obeying the usual constraints of "real" reality (Heim 1993; Larijani 1994).

Virtual reality by the grace of computer systems has some very practical applications, as in the simulators for training of pilots for airplanes and large ships. This can be done to great economy because such simulator training is less expensive and less hazardous than hands-on experience on real planes and ships. It can also be done without harm to human perceptions and mental habits because, in such applications, all the variables must be those of the real world, thus obedient to the toolbox of creation. Such applications have no use for designs made by incompetent architects, for instance.

In the more extreme applications, of arbitrary compositions of details, electronic virtual reality has the potential of doing severe damage to the human psyche, both individually and collectively (Rheingold 1991; Slouka 1995). Throwing overboard all the rules of nature certainly can lead some people badly astray. The toolbox of creation is compelling in the real world, but electronic virtual reality can disregard it, allowing many analogies of the house drawn by an incompetent architect to seem to be standing.

Amid all the fray over the accomplishments of electronic technology, it seems to have escaped attention many times that such technology really does not do anything new in principle, it merely sharpens and accelerates some things that were there all along. Virtual reality in quasi-logical laws is all around us and has done much harm ever since the codification of Roman law, and well before that. The money system works compellingly because it obeys the laws of arithmetic, but it also does harm by disregarding things of paramount value, by forcing a market price on everything.

LAWS OF NATURE ONCE AGAIN

Laws of nature are misunderstood if we think of nature (actual, existing physical, chemical, and biological nature) as being ruled by matter. In materialist conception there are usually but a few general laws of nature that might even be derived from a single overpowering primary law or "force"—the distinction between law and force is seldom explained. The immense variety of specific laws of nature, the whole of creation's toolbox, is then disregarded on the sweeping—and unfounded—assumption that all the specific laws might be derived from the general one if only we had enough scientific instruments to pursue our analysis to the bitter end.

We now propose the opposite to be true: Matter is ruled by the laws, the tools in creation's toolbox. These laws exist independently of time, space, and quantum (energy/matter), and they are indispensable for either of these dimensions of material existence to enter the sphere of embodied reality.

By way of certain types of virtual reality, such as money, for instance, the natural laws may reenter the scene where they are the least expected. Eventually, all real phenomena must obey the laws of nature. They exist out there, and not only in our minds.

Similarly, the very laws of life—vitalism's "will to live"—and "faith, hope, and love" are also preexisting principles always waiting in the wings wherever a planet turns out to be at least somewhat hospitable to life.

ULTIMATE CAUSALITY?

The classical conception of the world as ruled in every last detail by natural causality, or by unfailing divine will (as in the Book of Job), has been torn asunder by modern science, which insists on randomness and chance as backdrops to the actual unfolding of natural reality. Source of uncertainty, source of freedom—the perspective is hard to pin down in any definitive way.

In a probabilistic world, we can only foresee things in a statistical sense. If the weather report says that there is a 90 percent chance that it will rain in our area, then this means that it will rain on 90 percent of the localities, while every one-tenth will go without. The trouble is, we cannot tell or foresee which these 10 percent are. As long as the prospect is not 100 percent, I cannot be sure of rain on myself and my neighbors (the just and the unjust alike). That chance is, in fact, indeterminate.

Probability versus predictable causality—this was the basic source of disagreement around the quantum physics of the Copenhagen school

(Bohr, Heisenberg, and others). To the contrary, Einstein insisted that God does not play dice.

The trouble—and the fascination—with this question is that any definite answer is likely to be out of our reach. Probabilistic technology does deliver plenty of valuable practical solutions. This is even more so when classical probability is refined by the concepts and the analytical tools of fuzzy logic (Kosko 1993). For practical purposes, the world *is* probabilistic, not the least so in human affairs as we shall discuss later. We will get the most mileage out of any practical projects by using the tools of probability and fuzzy logic.

But even this does not have to be the ultimate answer. The ultimate answer may be practically out of our reach, but that does not have to render less important the distinction we are about to set forth.

The conflict just might be due simply to the limitations of our means of knowledge. For all we know, there might, in fact, be complete determination of every last little detail, but in ways that are too finely spun for us to perceive or even to formulate. The fluttering of a butterfly in Brazil just might be the ultimate cause of a thunderstorm in Chicago. If so, the logos becomes all encompassing.

When the instruments of knowledge forsake us, we are, in one sense, left with a margin for uncertainty, but also with a freedom to believe what we prefer to believe. Ultimate causality (Einstein) may be out of our reach and, if so, science cannot dictate our beliefs one way or the other.

EIGHT

Creation: The Ways Our World Was Made

iǫrð fannz æva / né upphiminn, / gap var ginnunga
(There was no Earth, nor Heaven above, there was a vast emptiness.)

Vǫluspá (Icelandic lore, ca. A.D. 1000)

The beginning of the world, as interpreted by recent science, seems to evoke less controversy than does evolution by creation. For one thing, of course, ultimate cosmology is still so speculative and presents so many alternative scenarios that most of us realize how remote all this is from our directly human concerns. Cosmological speculation has indeed gone to great lengths and in more than one direction (Hawking 1993; Parker 1988, 1990). In one sense, the biblical "let there be light" stands vindicated, for astronomers now tell us that the Cosmos consists mainly of light in all its forms—from gamma rays to radio waves. Interestingly, the theory of cosmic birth, which came to be labeled "Big Bang," was first proposed by an astronomer who was also a Catholic priest (Georges Lemaître).

In the end, of course, any coming-into-being from nothing remains as unexplained as creation by divine fiat. This ultimate reliance on mystery evokes less emotional resistance among modern people than does the attack on biblical authority coming from Darwin and other modern biologists.

Closer to ourselves, and less subject to doubts or alternative interpretations, are the modern findings on the coming-into-being of stars and planets from accumulating cosmic dust (Littman 1988; Allegre 1992; Sheehan 1992). From such results we also have fairly accurate estimates on the age of the Earth, at some 4.5 billion years, sharply at variance with biblical cosmogony (Dalrymple 1991). Planetary research has spilled over into geological history on Earth also because it helps understand past catastrophies of collisions with comets or asteroids, and the great species extinctions of the past (Raup 1991). Along with this we now also know much more about the dynamics of change in the Earth itself, with plate tectonics now firmly established (Erickson 1992; Officer and Page 1993). Biological life has also been revealed as a geological force, contributing to the changes in the ecosphere on which all living beings depend (Westbroek 1991). These very impressive research results seem to make very little dent into the convictions of those who continue to cling to a literal interpretation of biblical cosmogony, especially as regards the creation of mankind.

EVOLUTION VERSUS CREATION

When evolution replaced creation in modern thinking, something was lost in the bargain. The genesis of mankind was disengaged from the dramatic simplicity of biblical cosmogony, but the new doctrines were in a curious way made to explain themselves—a procedure that in the end, explains nothing. Evolution theory, as usually presented, explains how but not why.

The same conflict was experienced in ancient Greece, and one of the satirists dramatized it thus: Zeus was deposed by King Whirlwind: The old order, however irrational, was replaced by chaos, which is no less irrational. More recently, many writers have belabored the irrationality of existence. If it is all that irrational, from where comes the mind that is rational enough to be a yardstick? And if we did not make that claim for ourselves, how could we judge the rationality of the rest of the Cosmos?

The Darwinian dictum consists of two basic statements. One is that mankind evolved from animals as a part of general biological evolution. The other is that evolution was furthered by natural selection—"survival of the fittest."

The first point we need hardly discuss. Evidence is plentiful (S. J. Gould 1989). Only an obstinate will to remain ignorant succeeds here and there to render life difficult for an ordinary school teacher who only does his job. Evolution as such is fact, not just theory. But the second point is but a partial truth that has been badly overworked.

NATURAL SELECTION

There is no dispute over the importance of natural selection, which goes on all the time, for deciding which biological and human types will survive. The student of human affairs will add man-made, culturally induced selection of both plants, animals, and people. It goes on for thousands of years, shaping up cultural forms as well as many symbiotic and parasitic ones. But culture is but the prolongation of nature into a new dimension, so the contrast is apparent only.

Selection works with the biological stock that exists or comes into being by "mutations." In the latter we must now include also the results of natural genetic engineering, started by bacteria billions of years ago (Margulis and Sagan 1986). To natural selection, as defined, Darwin already added sexual selection, which leads to preference for certain characteristics of sex partners, especially the males. This has been pointed to as a complication also in recent research (Gould and Gould 1989).

Merciless toward the individual creature, selection works efficiently to reserve the future for stronger and more environmentally adjusted strains of biological and human life. As a concomitant of evolution, selection winnows and harnesses what evolution has brought out and makes it more orderly, viable, and inevitable. Sensitive to varying conditions, selection stands ready to draw the consequences and make the best of the situation.

But that is all that can be proven. There is no evidence to show that "survival of the fittest" has added any new features to those already present in the biological stock that is struggling and competing for survival. Before selection can do its work, something else must cause the mutations among which selection will act as arbiter.

A distinction is made between micro- and macroevolution. The former is readily credited to chance mutations. For macroevolution, which has produced major new features and major new lines of evolution, the origin in chance mutations may well be doubted. Recently it has been found that mutations occur routinely at a certain pace, measurable for each species—the so-called "biological clock." But this relates to all mutations, most of which are of no enduring consequence. For creative mutations, no clock can be established. Instead the concept of "punctuated equilibrium" has been proposed (Eldredge and Gould 1972 reprinted in Eldredge 1985).

Much has also been made of environmental adaptation. Many scholars have shown how the environment—in all its dimensions—may force a species to adapt. Much of this research work is admirable and its results leave a deep impression on our minds: How plastic and adaptable is living matter, to what lengths will it go to stay alive!

All of this, however, explains no more than does selection in general. Reactive adaptation can be shown to have modified many organs and properties of living beings, but it cannot be shown to have produced any of those biological innovations that made the organism more evolved. Many developments are explained by reaction to what comes from outside; but evolution in general remains unexplained.

Evolutionary progress from lower to higher levels of biology consists in the creation of more and more differentiated organisms. We may specify some major departures. First, there was the coming into earthly existence of bacteria, the simple prokaryotic cells. Whether this depended on spontaneous self-organizing within the so-called "primordial soup" of the early ocean or on seeding from outer space (the "panspermia" theory) we need not discuss here. For all we know, there may have been some of each. Recently we have learned (from Carl Woese) that the first to come along were the "archaebacteria," which lived without oxygen (and still do, as in botulism). Only after some long while came the second wave, the "eubacteria," which use oxygen—by then it had accumulated in the atmosphere as a waste product from the life of archaebacteria. Only after another good long while came the organizing of several prokaryotic cells into eukaryotic cells, the one-cell forerunner of multicellular biota. This stage already has bacterial symbionts as organelles within seemingly simple organisms, such as amoebas, for instance.

Multicellular life had a spectacular development some 600 million years ago—in the animal kingdom this broke out as the "Cambrian explosion." In a relatively short span of time (as evolutionary time goes) this phase created all the major divisions of animals known from later epochs (Gould 1989). Subsequent animal evolution apparently was all microevolution, elaborating on themes already present in the biota. From then on, it also appears that the main stems of animals have evolved as co-equal branches, not as "lower" and "higher" groups (Denton 1986).

Among the higher animals, evolutionary progress included the accumulation of larger and larger quantities of nerve tissue, organized into more and more versatile nerve channels capabale of receiving, processing, and storing more and more complex information about the surrounding world and about the organism itself. Writing off such events as mere "epiphenomena" is a kind of intellectual sleight-of-hand designed to divest oneself of the burden of explaining what cannot be reduced by mechanistic logic.

BLIND CHANCE?

Darwinian theory has recently been reaffirmed and expanded into a coherent theory of blind chance. Evolution could be read as if it were the work of a "blind watchmaker" (Dawkins 1989, 1995; Dennett 1995). The whole complicated series of events is declared to be the result of chance mutations accumulating over the ages. The fact that those mutations, which in some sense represent an improvement upon already existing genomes, are very rare—this is countered, in the doctrine of the blind watchmaker, by referring to the unimaginably long ages over which all this took place. Biological life on Earth dates back some 3.5 billion years, multicellular organisms about 600 million years.

The statement that all this was due to blind chance is, of course, just a supposition. It is the logical outflow of a mechanistic, reductionist view of reality. Occam's razor bids us to accept the simplest possible explanation, but there is no real logic to this, only a convenience for research work. Quite to the contrary, it has been held recently that the odds are crushingly against evolution by chance alone (Denton 1986). When rejection of the Darwinian theory leaves us without any theory of evolution—well, so be it. Recognized ignorance is better than any masquerade of spurious knowledge.

In the whole mechanistic approach, from Darwin to Dawkins, something essential was overlooked: the toolbox of creation. Sometimes it is termed a "grab bag" of nature (Denton 1986). Many of the tools in the box are synergisms, not to be reduced into simpler elements.

It is true that the content of this toolbox—the inventory of traits and constructs that could be viable in one environment or another—is very large. It is very large indeed, but it is completely dwarfed by all the potential trash that has no place in creation. Most of the combinations of elements that we could specify would not make viable compounds. Most of the combinations of known compounds that we could dream up would not make viable organs in viable organisms. It is evident that of all the spontaneous mutations that might take place by blind chance in a living genome, the vast majority are not viable; they are doomed to self-destruct, at once or within a short time.

There are more limitations than that, however. Not only would mutations by blind chance have to rely upon the viable potentials, the tools within creation's toolbox. Most of the time, even those viable potentials would not be welcome. To stay and to prosper, they would have to happen to come into being in a congenial environment, one where other viable traits make the new viable trait a possible addition to the preexisting genetic configuration. Most of the time, an existing organism would be such that most of the viable features that are possible because they are in creation's toolbox would still not be viable "in context." The opportunity for a chance mutation to set off a new evolution is limited to the cases where an existing organism has positive use for the new trait that has come in, or at least is not harmed by it. Mitochondria offer an extreme example.

The extreme instance where the tools of creation are exceptionally few and far between is the famous RNA-DNA sequence, the building block of genetic material (Shapiro 1991). It seems that there is only one such replicator in existence on Earth. Nearly all living things contain the same DNA, if in endlessly varying sequences (a few have only RNA, which comes close). No other known design of organic chemistry answers the requirement of a permanent, indestructible replicator to serve as building blocks of living genomes, not as far as the available evidence from our actual biosphere goes.

How could a blind watchmaker come across such a unique tool? How many hits and misses must take place in the "primordial soup" of preorganic chemistry from which life is supposed to have arisen? How probable is it that this would ever occur? There is simply no way of estimating here, no basis for accepting a blind watchmaker—other than by a self-blinded scientist.

SYNERGISMS IN BIOLOGY

Left out of the reductionist, mechanistic schema is all that this schema has failed to clarify. Foremost this relates to synergisms. These are rare in physics and are easily overlooked in inorganic chemistry. They are of overpowering consequence in biology. A great many biological processes cannot be explained purely in the terms of physics or elementary chemistry. Such cases in biology often appear as "black boxes"—one knows what goes into the process and what come out of it, but just how the outcomes are determined by what went in is often not at all explained. It is just stated, since repeated experiments give the same result. A natural invariance, as often as not, is simply its own explanation.

This applies very much to heredity. Before modern molecular science and the discovery of the RNA-DNA sequence, the existence of a certain gene could be said to explain the hereditary trait that depended upon that gene. This did not explain how the gene contrived to bring the result about. It was operational in crop and animal breeding, for instance, or for tracing hereditary diseases among people. Now comes the discovery of the DNA sequences, and the "gene" is replaced by such a sequence that is found to command the build-up of certain proteins. Explanation of heredity? Not quite yet. Explanation of protein chemistry, maybe, but heredity is still a black box. Proteins go in, and out come eyeballs and eardrums and stomachs and sex organs and all the rest. How do the proteins go about commanding body cells to specialize in such a way that an eyeball is created? This is still a black box—a more narrow one than before, but still a black box. It is not enough to state that cells in a certain position within the embryo or early fetus (the gastrula and its sequels) have some DNA sequences switched off, others switched on, to command them to form a certain organ such as an eyeball (Dawkins 1995). This is giving the black box even narrower walls, but the synergism involved still just is there.

To break open such a black box, one would have to be able to show how a synergism (in organic chemistry) leads to the creation of things that are qualitatively different from the sum of qualities in the proteins (for instance), and that can still only be done descriptively. Yes, this sequence of things going in leads to that outcome, but the connection is just another specific law of nature. It can be explored and described, but not really explained beyond the finding of an invariant process. It withstands all the efforts of reductionist science.

RECOMBINANT DNA

Into the whole theoretical scheme of evolution by blind chance there has recently been thrown in a new, rather unexpected train of observations. Recombinant DNA, recently the darling of manipulative biological technology, has turned out to be old hat in the laboratories of nature. From archaebacteria on, nature has played its own game of recombination of genetic material.

Credit for explaining this discovery belongs to Lynn Margulis (Margulis and Sagan 1986, 1991). The world of bacteria, which dominated most of the vast stretches of time when there was life on Earth, was a more or less continuous series of interacting organisms, with no clear species lines, continuously exchanging genes in a mysterious play of proto-sexual interaction. Such exchange of genes has continued up into the much shorter periods of multicellular organisms. Still today, bacteria and viruses serve as vectors for the transfer of DNA from one species to another. Here we are at a loss to say whether all the transfer of genes is really a game of blind chance or whether some degree of purpose might contribute to deciding which transfer of genes would be received as useful (or at least acceptable) components in the genome of the species to which the transfer took place. Nor is it clear how this process may relate to the biological clock of routine mutations.

An amateur entomologist's notebook may contribute some loose hypotheses here. The world of insects has cases of mimicry that may well be due to gene transfer. Moths that look like wasps are one case. An even more intriguing one is the large fly that is a nest parasite among bumblebees—it looks like a bumblebee, and it is tolerated by the bumblebees to enter their colony's nest. This really prompts the question whether recombinant DNA may not here have built a bridge that mere blind chance would be very unlikely to bring about.

The matter of nature's recombinant DNA is rendered even more intriguing because the same DNA can give different results as regards proteins commanded and organs formed—there is some latitude for variation in the language of nature (Margulis and Sagan 1986, p. 63). This comes close to posing the same question as formulated by Sheldrake: Are there patterns out there that may contribute to steering material nature?

Further ambiguity comes from the fact that, of the lengthy reams of DNA sequences in the genes of most organisms, a great deal is "passive," that is, has no known function in the heredity of the species. How does this play in recombination? The perspective is truly bewildering.

SYMBIOSIS

There has been a shift of emphasis in the explanations for evolution by recognition of the importance of symbiosis (Margulis and Sagan 1986). This has been labelled "evolution by association" (Sapp 1994). Symbiosis is an important modifier upon the general principle of competition, on which Darwinism draws heavily to explain evolution. A long history of research was largely neglected by the evolution theorists before the discovery of the RNA-DNA sequence. Symbiosis is, in fact, widespread both among plants and animals.

Even without active cooperation of species, there are many cases of subtle, if passive, interaction all over the place. For instance, the "hex-rings" of mushrooms, fleeing away from the soils recently polluted by the mushrooms' own decay products, leave ground to lush green grass thriving on the detritus of the receding mushrooms.

Among the more widely noted cases is that of lichens, for example, in the "tree moss" (Usnea) hanging down from trees, as in the lower Mississippi valley in Louisiana. Tropical rainforests abound in complicated systems of symbiosis as well as parasitic relations (Wilson 1992). Our own digestive tracts could not do their work without the help of intestinal bacteria, which in turn gain a livelihood in this manner. The same is equally true in the complicated digestive systems of ruminants. Likewise, the termites could not themselves digest the cellulose that is their food, but have preparatory digestion done for them by symbiotic spirochete bacteria in their intestines.

Symbiosis of people and domestic animals, and the special associations of domestic insects in our homes, have their parallels among insects, for instance. Ant hills house not only ants but also "ant guests" (myrmekophiles), some of which are useful to the ants, others merely tolerated. Other social insects have their analogues to these cases. Close cohabitation of species appears to facilitate the function of viruses and bacteria in transferring DNA—this is how the origin of several influenza strains in southern China has been explained as a consequence of the co-habitation of people and livestock.

ORGANELLES

The extreme cases of symbiosis as agent of evolution are in the instances where bacteria have become "organelles" (microorgans) in higher organisms. The green stuff that renders the botanical kingdom such a rich place for synthesizing organic material out of elements in the air and the water is due to *chloroplasts*, which really are green bacteria that long ago took up residence in other living cells and thereby rendered feasible much of the botanical evolution that followed. The chloroplasts, all along, are reproducing on their own (apace with the host organism), even retaining their own DNA separate from that of surrounding plant tissue. Such symbiosis is not what is usually meant by mutation, for the coevolution of chloroplasts and host plants is not necessarily dependent on any particular mutation in either. Obviously, such symbiosis must have changed the rules as to which mutations henceforth might be useful or tolerated.

The overall result of chloroplast symbiosis is impressive indeed. Just contemplate the high tree crowns in a forest—or even in the "urban forest" of alley trees—and understand that both the chloroplasts and the host plants have benefited hugely. Both of them have gained much larger scope for quantitative growth. The taller the tree, the more layers of air can be drawn upon for carbon to go into photosynthesis. Trees produce more biomass per area unit than do any herbaceous plants in the same environment. Climbers, as tree parasites, get the best of two worlds.

Another parallel, at least as impressive if less easy to visualize, is in the mitochondria, which co-exist in the body cells of plants and animals, ourselves included, and including also the advanced one-cell species (the eukaryotic protists). Without the mitochondria, our bodies could not perform the basic oxygen metabolism on which all multicellular life on Earth depends. Again, the mitochondria retain their own DNA and reproduce in a process that runs parallel with that of the host organism. In sexually reproducing species, this is usually only through the mother side of the family tree—human sperm cells, for instance, do not contribute any mitochondria, only egg cells do.

It appears that the DNA of mitochondria mutates more frequently than that of the host organism. Yet there has not been much evolution of these elementary organisms over the 1,000 million years (or so) that mitochondria have been combined with organisms that are more complex than themselves. Some forms of cancer might depend on mitochondrial mutations, but there has been hardly any scope for such organelles to become much more evolved than they were at the beginning of their term of service. This appears to be an instructive case of overarching constraint, rendering most mutations unwelcome "in context."

A possible third case, which is less well understood and may not yet be completely clear, is in the *undulipodia* ("waving feet"), which play a vital role as cilia within some organs and as a motion mechanism, for example, in sperm cells. It has even been suggested that the cells of the human brain (and many other brains) could be understood as transformed versions of the same microbe that might underlie the existence of the undulipodia (Margulis and Sagan 1986). This is a class of bacteria known as spirochetes, extremely mobile microbes, some of which can be very dangerous (e.g., the syphilis microbe).

PROGRESS BY CHANCE?

There is no particular reason why the achievement of greater organ differentiation should lead to any superiority in the struggle for survival. Some of the oldest and simplest forms of life are also among the most durable ones, having survived both the evolution and the extinction of far more sophisticated forms dating from more recent epochs.

Bacteria are the most elementary of all known organisms. They are also the most surviving, both in their independent existence as nature's recyclers and scavengers, and in roles as symbionts and organelles. Some rather primitive life forms, such as the amoeba, are first-rate survival propositions. The giant squid of the ocean depths is a very archaic creature, with huge aggregate body weight; as a survival proposition it no doubt exceeds many of the "higher" species. Sharks of the high seas are more recent, but some at least are very long-lived species. The opossum, a permanent resident of North America, is also a very durable creature.

The so-called "biological clock" of a constant rate of mutations leaves us without answers here. Evidently, there are different constraints as to which mutations will survive.

If environment adaptation was all that counted for evolution, then a few equilibrated simplicities might have continued to reproduce over the aeons—if they had been created at all. Migration could have answered most of the adjustment problems caused by climate change, for instance.

How little of the progressive organ differentiation really served the purpose of survival is perhaps most evident in the case of the human species. Its genesis, by way of anthropoid ancestors, implied above all the creation of the human brain. This wonderfully sophisticated instrument is estimated to include some 100,000 million neurons, all in complex interaction with each other and capable of unspeakable degrees of data synthesis. It seems that our brains represent a rather extreme case of excess capacity. We are only now beginning to tap the riches of this instrument in a systematic way, while admiring its spontaneous performances all the way from Altamira to Michelangelo.

What was the use of this precious instrument in the jungle? Could primeval man survive better than the monkey? Nothing indicates it. Both of them lived in a monkey's world, and the monkey was perfectly capable of handling the situation. Before the coming of modern technology, the baboons of Africa may have been a greater biological success than Africa's Homo sapiens (Strum 1987).

As for survival, it can be discussed whether insects are not really a more powerful presence on Earth than are all the prehuman "higher" animals taken together. Insect colonies function in some ways like superorganisms.

It is quite plausible that a new organ differentiation was rather a handicap before its carriers achieved full adaptation to their new equipment. The higher animals are on the whole—in the natural state of things—present in smaller numbers, with smaller aggregate body weight, than the simpler, "lower" ones. Usually they have lower fertility as well as a shorter life expectancy, as species, before extinction.

Higher organ differentiation means more specific requirements on the environment, hence a more narrow range of possibilities to live and more vulnerability to hardship. Is this an aid to survival? It looks like the contrary. Only subsequently can the more evolved organism come into play by more cunning and more versatility, but such adaptation comes after evolution, and it is doubtful whether it achieves more than a compensation for the handicaps of differentiation. Human history is the only strong case to the contrary, and it is essentially an affair after the decisive biological evolution had been achieved. The advanced phase of mankind's power over nature is also recent enough and unbalanced enough so that, in the long sweep of natural evolution, it may be uncertain whether it is here to stay.

The slowness of mankind's path toward cultivating its marvelous brain is a good example of the "higher" organisms having heavier odds against them. Their creation is not to be explained by simple reference to mechanical or passive reaction to an environment. The hominid stem became an ecological dominant by the conquest of fire, and that innovation cannot be plausibly tied to any particular level of brain development.

This leads us to contemplate the coming into existence of organic life as a whole. It represented a huge "leap forward" in the direction of increased complexity. This would seem to fly in the face of another main direction in the evolving of the physical cosmos. Physicists have formulated the so-called law of "heat death," or the downhill movement of physico-chemical complexity. Disaggregation into simpler elements, releasing energy in ever simpler forms, is the normal thing in the physical Cosmos. Aggregation, build-up toward higher complexity, is described as an "uphill movement" and is rated exceptional in the physical Cosmos. Some of it goes on inside the stars, creating heavy elements out of lighter ones, but this is supposed to be always at the expense of energy, so that the higher complexity generated is of lesser volume than the simpler one replaced.

Biological genesis is the extreme case of the "uphill movement," the unlikely thing in the world of physics; and evolution is the continuation of that uphill movement. The well-worn "instinct of self-preservation" has been reformulated as *autopoiesis*, the tendency of living matter to repair and reconstitute itself. Eventually this leads to the ability to replicate indefinitely, as in the microcosm of bacteria and in the RNA-DNA sequences of living organisms.

VITALISM

The quest for an ultimate cause led to the assumption of a "vital force" in the thinking of Lamarck and Bergson and many others. Classical vitalism is now generally rejected, not just because the presence of a vital force cannot be proven, but rather because the assumption appears unnecessary. The whole process can be described, nay, "explained," without such an assumption.

This has not deterred some more recent biologists from musing over these things and sometimes they formulate what amounts to a philosophy, or a philosophical hypothesis, on living matter. Its essence, they say, is the ability to react actively: to receive and emit signals and to evaluate them emotionally. This ability, some speculate, may also be present in inorganic matter, although in an extremely much weaker degree. Even the simplest organic molecule is so enormously more complex than the most complex molecule of inorganic matter, that this alone would be sufficient to explain the apparent gulf between the inorganic and the living.

By this line of thought, living matter is the more living, the more it is capable of doing the things just mentioned—that is, the more it is organically differentiated. Since the highest degree of organ differentiation is that present in the species with the largest and most evolved central nervous system, this would place the large-brain mammals at the apex of evolution—except, of course, for the handicap of extremely large body size, which appears to be keeping elephants and whales from becoming as evolved as mankind.

A complication arises also from the consequences of social organization. Ants, termites, and other social insects may be organized to the point where the colony becomes a superorganism, capable of thinking farther than any of the individual members. This may be among the areas where we are systemically incapable of following paths other than our own—animal communication certainly is, to a large extent, incompatible with our mental equipment.

From neovitalism, we are led to formulate a belief on the nature of biological genesis. If matter is all alive in some degree, and if this aliveness is enhanced by more complex and more centralized organization into beings with higher organ differentiation, then there might well exist an inherent tendency of matter to achieve such organization, a "will to live" that comes out spontaneously whenever there are no insurmountable obstacles (as there are in most of the stellar Cosmos) (Coveney and Highfield 1990). The rareness of life, and the fact that on our planet it apparently was created only at some unknown time or times in the remote past, would then depend on the necessity to have an unusual set of favorable circumstances in order for the process not to be practically impossible. Something similar might apply to the various breakthroughs toward higher forms of life: They, too, may well have required particular circumstances in order to be feasible.

This formula about an inherent will to live, or a creative urge in the Cosmos, does not amount to an explanation in the scientific sense. It is just a belief—or, indeed, no more than an alternative wording of some trivial experience. But it is better to have such a belief or to use such a wording—all along knowing what is involved in this kind of assumption—than to carry on the illusion that mere reference to descriptive detail, to blind chance, or to mechanical forces could explain the constructive change from simple protoplasm to brain and other tissues. From a scientific standpoint, the formulation just proposed amounts merely to a confession of ignorance. But again there is nothing unacceptable about such a confession: It is definitely better than the semblance of knowledge where none exists.

Evidently our knowledge has to stop someplace. We have to have axioms and other basic assumptions that cannot be proven; if they could, it would have to be by means of still more elementary axioms, and so on. Such unproven assumptions are the rock bottom of our knowledge. A piece of confessed ignorance can also serve as such a rock bottom, whether in time it becomes recognized as an axiom or not.

The flaw in the mechanistic, reductionist view of biological reality comes from the limitations on our instruments of knowledge, especially the inability to reduce synergies to their components (as we have discussed more than once). What is known for sure is what could be made known by using the instruments of conventional science. All the received truth is steeped in mechanistic terms, for the simple reason that whatever could not be clarified in such terms has remained more or less unknown. Dogmatic reductionism operates a set of logical circles.

It can be discussed whether a will-to-live assumption is any more arbitrary than are the assumptions of basic forces in physics. The latter are allowed to become articulate as entities in algebraic equations, while the will-to-live is not so used. We will retain it as long as it does not come into conflict with well-known data. It falls in line with recent thinking on "order out of chaos" and the tendency for self-organization.

This reasoning introduces again creation on the scene—self-evident and mysterious in its transcendent majesty.

CREATION

By "creation" we intend to designate the entire uphill movement of biological evolution and human development: the producing of more and more complex designs and processes from simpler elements, in defiance of the apparent general tendency of matter and energy to travel downhill by way of disaggregation and disintegration toward simpler conditions. We could include in creation also the "cooking" of heavy elements in the stars, or even the hypothetical Big Bang itself, as belonging in the concept of creation (Davies 1988, 1992).

Reproduction of living life, on the other hand, is not in itself creation in this sense. It may amount to nothing more than maintaining a steady state, albeit an extremely complex one, existing in an incredibly fragile equilibrium of mutually countervailing processes both constructive and destructive. The replicators (RNA/DNA) are part of such a steady-state scheme, and so are the proposed new cultural replicators, the "meme"s (Dawkins) or "seme"s (Margulis) (cf. Plotkin 1944).

Creation, by contrast, is a chain of events that step by step breaks the barriers of existing equilibria. It may be accompanied by destruction that wreaks havoc with much of what previous creation has brought into being. As long as a creative innovation of higher complexity survives well enough to reproduce, it remains a dynamic addition to reality and becomes a building block for a cosmos of higher complexity of designs and processes than the previous one, hence more intensive existence than before.

We are not discussing here the origin of matter and energy in our cosmos, or of the physical Cosmos as such. We do not even have to choose between leading theories for the origin of life on Earth—the primordial soup versus panspermia. The whole chain of bio-psycho-sociological creation is more than enough to contemplate at one time. Photosynthesis, cellular organization of living matter, symbiosis leading to organelles and multicellular species—all of this forms a chain of uphill movement, in defiance of the general drift toward entropy.

Mankind is not the end of creation. Quite to the contrary, we are an instrument for continued creation. Biological evolution is continued through the sphere of the human psyche into the creative realm of human culture. We were created by evolution, and we continue to create through development. On the threshold between the two, there are little-understood processes that may need some discussion.

DOMESTICATION

The biological path from hominid to mankind may have been part evolution, part domestication. We do not really know how this phase of biological change proceeded (Leakey and Lewin 1992; Howells 1993). There are loose hypotheses around, for instance, about the disappearance of Neanderthal man: This type is generally supposed to have ceased to exist because it was displaced and made extinct by the expansion of more highly equipped competitors (Shreeve 1995). But who knows, maybe Neanderthal man was not displaced but absorbed? As a wild hypothesis, let us try the notion that the Neanderthals were red-haired. (Remember that red hair is *not* a variant of blond; it is a different hair type altogether.) If so, could the occurrence of red hair among the peoples of Europe and the Middle East signify traces of Neanderthal genome? Such a hypothesis can neither be proved nor disproved, but the suggestion may serve to indicate the width of our ignorance.

Somewhat less forbidding is the question: Is mankind a domesticated animal—domesticated already in the biological sense? Domestication of plants and animals into culture forms (cultivars, cultigens) is much more than just the taming of wild species. To a great extent it also has implied true biological change. Among the simplest cases are those of some standard food crops, such as wheat, corn (maize), and rice. None of the high-yielding strains is identical with any wild species. The ordinary wheat plant (bread wheat) has been found to be the result of crossing three wild grass species: Of its 42 chromosomes, only 14 belong in the wheat family. For corn, the ancestry is less well-known, but it is certain to be an interspecies cross, with twice the chromosomes of any wild ancestor.

Even when a cultivated species or variety is merely the result of biological change through mutation or polyploidism (multiple sets of chromosomes), the effect is usually the same as with interspecies crosses: The new forms that are especially useful to a human cultivator are not viable in wild life. The forms useful for people are the high-yielding ones, producing much fruit and proportionately less foliage. In the wilderness, they would succumb in competition with plants that develop more foliage and less fruits—the weeds, as we call them. The weeds are fruitful enough to replenish the Earth without nourishing very many people in the process.

Under cultivation, the rules of competition are reversed. The forms that are low-yielding in human food are systematically removed (by weeding), while those high in food yield are made to survive and to multiply. There is more: Among the crops, the most high-yielding plant specimens will contribute the most to next year's sowing seed. This would boost the general yield level automatically, even without any conscious purpose.

There are analogous processes about domestic animals. The culture forms are all different from their wild ancestors, some of them radically so, even without any interspecies crosses (of which there are some). The selection of hybrids to preserve and to multiply (and eventually to pure breed) took sight both of productivity and docility (in dogs, these qualities shade over into each other). The central fact is that the shaping up of these culture forms took place in an environment that had been modified by mankind as an ecological dominant—a creature able to modify the rules of competition as they had existed in the wilderness.

This is why we should also contemplate the possibility that Homo sapiens as we know the species may be the result of changes in the direction of evolution that were determined by the incipient ability to master nature and change the environment.

Some of the hominid ancestors may already have used fire and simple tools. Changing the rules of competition may have contributed to permitting more highly brained mongrels to survive. As in the domestication of animals, captured enemies ("aliens") may have contributed to the mixtures. It is true that in the most primitive human societies, aliens were as alien as animals. But it is also true that in such societies people lived closer to animals than in more civilized times. The cases for mongrelization may have been far between, but for that sake they need not have been few, for the ages over which these prepaleolithic and early paleolithic contacts took place were so much longer than all that followed. Even long before the coming of classical slavery, primitive tribes may have taken the women of defeated tribes and made them part of the victorious tribe. Warfare was not unknown in those remote ages. There is evidence of violence between people, from Neanderthal remains some 60,000 years old.

Human mongrelization is likely to have contributed to survival by more biodiversity. As more types were created in this way, more of them could be tolerated and the more differentiated society became. Human society tolerates inadequate individuals as long as they are useful in specialized tasks. For instance, ancient myths picture the God of metallurgy as a cripple. Thus selection has not been merely of the "best" in the sense of excellency in each individual, but also in the direction of making the population more versatile. Where nature has "ecological niches" defined by physical ambient, culture offers "survival niches" defined by specialized activity.

One conclusion is that "natural man" may never have existed. If so, we cannot backtrack to such an ancestral type even if we wanted to because there may be nothing to backtrack to. The prototypes may be lost.

Creation still goes on. We cannot stop it without destroying it. The question is: Can people "control" creation from now on, thus shoving aside and replacing whatever processes that steered it from the first spark of elemental life up to the present?

The wonders of science have inspired some people to think that from now on, mankind will be practically free to do whatever it likes—to project into reality any whim or urge that people may experience. This, of course, still raises the philosophical question: From where do these whims and urges stem? If they come from the original processes that created us, then we may not be as free as some people believe. But leaving that aside, can people create past themselves? And if so, what would be the meaning of this?

CREATION BY SCIENCE?

One of the boldest assumptions made in the scientific establishment is the statement that people eventually will be able to create life in test tubes, "from scratch." Finally we might not only duplicate the human species in a laboratory, but also produce even "better" synthetic beings— supermen to replace ourselves and bring mastery over nature to perfection. One writer not long ago used the quasi-romantic expression that such a synthetic, superintelligent being will have been "conceived in the love-bed of science."

Such a futuristic vision badly risks overstretching the capability even of human creativeness. Whatever is invented must obey the rules of creation's toolbox, including the array of synergisms that are allowed within that toolbox—a tall order.

Let us quote the analogy of experiments with computer produced music. Computers can manipulate what musicians have already come up with, but if they ever exceeded (or simply matched) the creativity of great musicians, we do not hear much about that. It could happen by random chance on occasions few and far between. Instead of having someone sit through all the reams of worthless machine compositions, it is much more practical to spot budding musical geniuses and let them write good music straight away. Genuine creativeness requires mental powers about which we do not know. We must reckon with new synergies, not reducible by any mathematics or logic.

This is why the talk about building supermen in laboratories is likely to be sheer hubris. Biologists will create better strains of crops and domestic animals. Maybe they will eventually be able to produce entirely new ones, even though it appears somewhat more practical to do so by manipulating the genetic stock that is already here, rather than going to the awesome task of constructing genetic material from scratch by synthetic processes. All of this manipulation of other forms of life is, in principle, no more mysterious than the manipulation of dead matter to build ploughs and shoe factories and computers. All of these things can be done because they are projections of the human mind, which reaches out and realizes some part of its potential. In so doing, we take the central purpose of human existence for granted. All the things that people can master are simpler, of smaller design, than the human mind itself. But in duplicating a human being, let alone exceeding one, we would endeavor to create something the same as, or in excess of, our own size in design, and that must be seriously doubted if it is possible. It could at most be an unintended outcome, by random chance, unpredictable and fraught with unknown dangers, like the classical image of Frankenstein.

Any other task, whether of creation or of routine, can be handled. It is, in principle, possible to channel through a human mind anything that is within its own capacity. The only thing not within the processing capacity of a human mind is another human mind in its entirety. This is as impossible as it is to lift oneself by the hair. Even if a team of scientists were to pool their talents to build a synthetic superman, this creature would lack character, hence mental viability. Like the dull computer symphony, he would appeal to no one, probably not even to himself. Some degree of self-esteem is indispensable for living—lack of it leads to suicide. To create a whole human character, the way a musician creates a symphony, is too large a task for any scientist or scientific outfit. For creation to exceed the present mankind, we can only hope for the creativity of nature itself.

Since art is an expression of life and not just mechanics, we also must realize that each living species is a work of art. Mankind, with all its cultural accomplishments, is the greatest work of art of them all and the only one that, by definition, we are not equipped to duplicate, let alone exceed.

The irreducible conclusion is that we can master all else, but not really ourselves. The Earth may not be our fate; we can manipulate it and make it produce. Heaven and stars may not be out of reach; given time, someone may get there—for whatever purpose that is desired. Our physical nature can also be controlled up to some point that may be farther away than we think. Birth control will eventually give us a "people-o-stat" to set the ideal number of our species to be in existence at any given juncture. All that people can touch is theirs; only our own soul does not belong to us. Tinkering with the properties of the mind can destroy it, but in a positive direction we can only discover what it is, not really improve upon it beyond a full realization of its age-old, unfulfilled potential. All of our motivations reflect what manner of being we became through creation. All our creative and productive avenues will only lead to projecting into reality the things that were made possible—and maybe inevitable—because of the way we were created.

INTELLIGENT COSMOS?

In a finalistic setting, the *logos* of the human soul is the ultimate *cause* of all the more or less "material" stages through which creation had to proceed, logically building up the more complex from the less complex. One of the least conformist among the astronomers, Fred Hoyle in England, has suggested that the Cosmos developed toward some preset goals (Hoyle 1983). The totality of all the logos would then have functioned as if it were a cosmic intelligence.

The logos being preexisting, it has really, all the time, ruled over the atoms and the molecules, the body tissues, and the brain cells. The organizing principles were there, prior to all that was organized at their command.

A construct of this encompassing magnitude cannot be subjected to any scientific test to prove or disprove it. Nor can there be any proof to the contrary. None is possible any more than there is any proof positive of the Big Bang or of any other ultimate cosmological origin we may think of. But Hoyle has given us something to think about; it is worth pondering.

NINE

History: Limits of Science in Human Affairs

Il n'y a qu'une science de l'homme dans le temps
(There is only one science on mankind through time)
                    Marc Bloch, *Apologie pour l'histoire*, p. 15.

The need to know history is now more compelling than most modern people seem to realize. The lesson was taught forcefully by Orwell in *1984*: Falsified history can falsify our lives altogether. The words of Marc Bloch give a healthy breadth to the ambitions of scientific history. There are lessons to be learned, but what are they? Do they come from history as a web of past facts or from history as a narrative trying to cover those facts?

In the short term, history may seem to teach lessons of what to do and what not to do, given some stated political goal. Lenin thought so when he said that "revolution teaches" (meaning the events of 1905–1906). He may or may not have read the nestor of Russian historians, V. O. Klyuchevsky, who looked upon history as a *magistra vitae* in the sense of a supervisor: Instead of teaching lessons, she punishes those who do not heed those lessons. Orwell might have agreed.

The need is not always matched by opportunities. When "man is the measure of all things," it is difficult to make us a measure of ourselves. The sciences about Homo sapiens as mental beings are obviously the least perfect of all sciences. Their annals contain a depressing amount of spectacular failure as well as a bewildering degree of general untidiness. False starts and faulty orientation lead to backtracking, which is all the more tiresome if one walks in circles. The defects which are so apparent in the social sciences and the humanities are, however, not all due to the classical difficulty of knowing ourselves. If these defects amounted to no more than the cumulative reflection of individual and collective conceit, self-deception and dishonesty, they could, in principle, be tackled by some ingenious system of cross-checking—between individuals, between groups and classes, between nations, and between generations.

Much of the difficulty instead derives from inherent problems of the human condition itself: Chaos, contingency, and fuzziness impinge more severely than anywhere else in known existence. This is more true the more the historian's ambition is directed toward "history as a whole." This vast concept is too unique, too chaotic, to be analyzed by scientific method—it can, at best, only be described. Arnold Toynbee almost threw up his hands because history appeared to be just "one damned thing after another." Partial processes of technology, economy, culture, and so on are more likely to be accessible by scientific analysis.

Before elaborating on this vast general theme, let us retrench and show that the humanities—history in the wide sense—have more hard-core scientific stuff than is generally recognized by natural scientists trying to critique historians.

"SOFT" SCIENCE?

In the attempts at a dialog between the natural sciences and the cultural sciences, it often appears as if there could be more than one concept of science. Many humanists are people of great erudition, so what is wrong with them as scientists? On close inspection it turns out that the sciences about Homo sapiens as a cultural artifact contain elements where science can be as hard as in physics or chemistry. The difficulty begins when such elements are to be spliced together, using more or less speculative statements on the wider reaches of the human phenomenon, those where contingency begin to take over as a leit motif.

Hard science in history is all too obvious if you look in the right places. Coming as I do from the economic and social history of rural Europe in the Middle Ages and early modern times, I know mountains of research results that are as rock solid as any from experimental physics or chemistry. The hard core in the historical sciences is in the study of archives, the written remains of human life as it was once lived and endured.

Knowledge of documents, as testimony (really, remains) of past human activity, is in no way inferior to the knowledge of rocks and fossils on which geologists and paleontologists base their constructs of land-scapes created and destroyed or of biota evolved and extinguished.

The hard science of historical documents began with Jean Mabillon (late 1600s), who invented a means of telling apart genuine documents (those that are what they purport to be) from forgeries. His evidence was based on a version of what we now recognize as "informal statistics." If handled correctly, informal statistics can be as convincing as any other kind of scientific inference. This analytical tool may appear crude when compared with the statistical inventions now routinely applied in natural sciences. But in archive science the elementary nature of the analysis is helped by the often massive sets of documents to which the analysis is applied.

In my old field of agrarian history, the results of document study can often be strengthened by looking at the landscape itself. It is permanent, or nearly so, on the time scales where people are living and historians are working. Past information about land areas and their legal status and their economic value can often be checked against the physical facts of the very soil where the rights and the economic returns were played out (F. Dovring 1951, 1953).

There are parallels in many other lines of historical research. For instance, in-depth studies of special human societies, such as dissenting religious groups, can reach high levels of scientific certainty in many research results. Many such results are rendered solid by the invariances that can be observed among the actors in typical conflicts and among the expressions of their faith as well as their personal stance in typical conflict syndromes (K. Dovring 1951, 1954–1955).

CRITIQUE OF SOURCES

Authenticity is only the first problem about sources. Critical appraisal is often a long and difficult process. It begins by distinguishing between vestiges and narratives.

For instance, a letter, whether private or public, is a vestige of some activity on the part of the letter writer. If the letter can be proven genuine, the contents not only testify to some activity of its writer. It may also confirm or reject some assumptions as to what happened or might happen.

At the same time, the letter may tell some story. Then there is nothing that can be said a priori about truth or falsehood. Errors of fact and judgment can be intentional or accidental.

Critique of sources belongs not only in the appraisal of written remains of the past, it also applies to contemporary documents. For instance, official statistics are, for the most part, narratives; they tell some story for the sake of which the data was collected. The vestige aspect is usually minor. Critique of such sources demands not only statistical criteria of the quantitative data. It also requires appraisal of the people who made the sources. Recent revelations about the FBI have shown that its official truth often cannot be accepted in a court of history.

The problems of source critique can only be sharpened by the increased speed of transmission and the increased access. Take the Internet. How many alternative appraisals of sources can there be room for here?

CONTINGENCY

The most powerful barrier against full historical insight is in the complexity and variability of the human phenomenon. The problem is often shunned and seldom acknowledged in full. Many humanists and social scientists will admit its existence in principle, in a noncommittal and indeterminate way. Some will blame it for lack of statistical significance in their data. A formal measure might gauge the scope of this variability for each specialized subject, dialectically isolated from its setting. For the totality of human existence, this variability is the product, rather than the sum, of those for special subjects. We should apply multiplication, nay, exponential expression, rather than addition here. Before we try to interpret the whole shimmering, evasive panorama of human events, we should have some notion of the chance that we may succeed. There is a sobering numerical exercise that should precede any positive statements of what can be done.

Just how complex is the human phenomenon? Inorganic matter is infinitely simpler, trite as that may sound. Yet no one finds physics or chemistry or astronomy to be simple subjects. Living matter, even when seen merely from the physical side, is so enormously more complex than inorganic matter that the biological sciences must put up with considerably more generalization, and with less absolute preciseness, than physics or inorganic chemistry. When it comes to the prolongation of biology, by way of the nervous system into mental designs and their ramifications across all fields of human action and interaction, then the degree of variability defies ever so tentative gauging. All of the 6 billion (or so) people in existence today are different to look at, and the amount of difference that is housed behind our faces is usually much larger than can be judged from the outside. Not even identical twins have the same fingerprints, and only people who know them superficially can believe that they always think alike.

GENETIC VARIABILITY

The least complex part of the story is in fact in the biological variability of the human species. Independently inheritable genetic factors are estimated in the thousands. Genes may also have a multiple incidence on several individual characteristics. Taking the simplest case, each gene (DNA sequence) has at least a significant effect on one characteristic with at least two alternatives (such as blue versus brown eyes, for instance).

Thus each pair of genes allows at least two genetic combinations and sometimes more. Taking two as a minimum, we can designate the number of gene pairs as x, and then the amount of possible genetic variation would be the xth power of 2 (2 exp x). With the value of x in the thousands, the xth power of 2 is an enormous number that cannot really be imagined. The tenth power of 2 is 1,024, and the eight-thousandth power of 2 is a number with upward of 2,400 digits.

The entire number of people existing on Earth at present is somewhere near 6 billion. This is only a ten-digit number. The total number of those who have gone before us has been a number of eleven digits or so. If mankind were to grow a thousand times more numerous than it is now (a rather absurd proposition, materially), the number would still have only thirteen digits. Repeat that over a million generations, and you would still have only a thirty digit number (about the sixtieth power of 2).

What a far cry this is from tapping the immense genetic variability of our species. Technically there are short-cuts that might reduce the scope of this variability, such as systematic purebreeding and the recent scare about clone formation by vegetative reproduction, but these we have no reason to discuss here. Short of any such artificial impoverishment of the species' biological treasury, the chance of repeating identically the same genetic setup in two different individuals is so close to nil that we can stop wondering why we are all different.

By far the largest number of possible human characters—in the genetic sense—never get a chance. Each of us comes into being through a gamble in which the odds are crushingly against us. Why am I here and not a brother or a sister who might as well be more different from me than I am from the relatives I have actually known? The bottomless space of random chance gives no clue. Repeat the same question in regard to each and every one of our ancestors, and any positive chance seems to disappear as if in the vastness of the astronomic cosmos. Worse: The number of atoms in the known cosmos has been calculated (by Paul Dirac) to be a number with a mere seventy-eight digits.

And yet, all of this genetic variability is dwarfed when compared with the variability that comes from learning and shaping up after we were born.

CULTURAL VARIATION

The register of possible environment imprints upon the plastic substance of our central nervous system is wide to a degree that is even more beyond comprehension than that of genetic factors. Whether these possible imprints are in themselves more numerous than the genetic factors—as they most likely are—is not even the most important thing here. What compounds their variability is the fact that experiences can happen in different time sequences, giving rise to even more synergistic combinations. The mathematical expression here includes something called "faculty," which gauges the amount of combinations from which reality takes its pick.

Faculty is the number of different sequence-combinations that can occur with the number of variables at hand. Faculty 1 is 1, faculty 2 is 2, faculty 3 is 6, faculty 4 is 24, and so on; faculty 15 is already over 1 million million (twelve digits). When the number of variables is increased, the number of possible sequence-combinations snowballs at a fantastic rate. Each added variable over 10 will add one digit, each variable over 100 at least two digits, and so on.

The fact that not all environment variables can occur in all possible time sequences is offset by the very large number of environmental factors that can apply to one human experience. The number is rather indeterminate and changes over time as human culture and other relevant circumstances change. All told, the number of possible time sequences of environmental factors will amount to a number of several thousand digits—to serve as exponents of 2. The individual will thus be one of a number that is the several-thousandth-digit-number's power of 2. The number of possibly occurring human characters—in the full sense of the expression—is thus an astronomical multiple of the possible number of genetic combinations, which, as we saw, is already vast beyond comprehension.

All these numbers are, of course, such that it would be futile to even compute them, let alone try to visualize them. Holding this under the reader's nose would be merely facetious were it not for the imperative necessity to rub in, until we never forget, that the human species is subject to random variation to a degree that defies imagination no less than it defies any possibility of exact computation or even approximate estimation. The statistical "law of small numbers" throws a heavy shadow over the universe of humans, far more than over any other universe in the Cosmos. Random effect is a source of disorder that constantly walks at our side. Seen from another angle, this is what makes freedom possible where an accurate clockwork of blind laws of nature might render all outcomes predictable.

It is important to accept uncertainty. Try to compare law and history. In a court of justice, everyone is innocent until proven guilty. In the court of history, everyone is suspect until proven clean.

RANDOMNESS

The human experience of randomness has been expressed many times, perhaps never as forcefully or as pessimistically as by the existentialists. Jean-Paul Sartre (in *La Nausée*), contemplating the meaninglessness of existence, chose a trivial object—a tree root that bulged through the surface of the soil in a public park. The cultivated specimen of a well-known botanical species, growing in a civilized place, displayed innumerable traits distinguishing it from all other roots even of the same tree. There was no rationale as to why it had just those bends to the left or to the right, just those individual fissures, and so on. It just was there; it just existed. It might as well not have been there; something else might have existed in its place. When this happened to the simple things, what about the human individual? Each is obviously the victim of sheer chance, and any idea of security or rationality in this random existence must be a cruel illusion.

The emotional mood pictured in Sartre's *La Nausée* reflects the predicament of an individual who has lost his bearings. All the values he once believed in seem to have failed, and in the vacuum before new ones— or old ones, reinterpreted—have taken shape to organize him afresh, the randomness of the human condition is felt all the more acutely. E. M. Forster in the 1920s (in *A Passage to India*) met the same problem when confronted with the echo chambers of the Jainists. In more normal circumstances, our organized personal motivations appear to fall in line with a general purposiveness of our cultural, no less than our psychological and biological, character.

That kind of situation is, of course, present, consciously or not, in all of us as far as we have been able to pursue goals of any kind. To some degree it is necessary if we are to live for any prolonged time. At the same time, this spontaneous purposefulness is a powerful source of misjudgment. Purposive individuals tend to take their own experience and their own value judgments for being more universally valid than they are, and to overlook the diversity abroad in the value judgments, as well as in the experiences, of other people. The existentialist in *La Nausée* made more of life's randomness than it deserved. The purposive individual, in unconscious or unreflected striving for conquest and transformation, underrates the width of diversity that we are all up against. Willpower and wishful thinking are Siamese twins. Here is also the main wellspring of pseudoscience in human affairs.

HUMAN NATURE

This is also why many people have far too decided ideas of what "human nature" is like. As an opening statement we submit that human nature is a good deal less uniform than it appears to the naïve and the narrowminded. There is much more diversity and disorder in the human universe than most of us can stand, emotionally, in our craving for a set order to rely upon. Too many have tried to ignore diversity as it exists when they attempted to impose their own character and their own intentions upon the world. This kind of oversight is so common because of the limits to communication between people: When we meet strangers, it is the common ground that permits any communication at all. The differences are often such that they defy communication, and when they therefore remain inarticulate, they may be underrated entirely. The more we know others, the more we recognize how different they are from ourselves. A next-door neighbor is often found to be different to a degree that many of us do not admit about complete strangers.

The surest path to knowledge about people still goes over the tabula rasa of the old man in Athens (Socrates): Know that you know nothing. From a sure-fire belief in some elaborate system of "world explanation," you will only be gradually pushed down the slope of disappointment when the old assumptions are failing bit by bit. From the abyss of nausea, of despair over the general disorder, a new freedom from expectations will allow us to rise refreshed and contemplate some of the basic order, beauty, and meaning of the Cosmos. They were there all the time.

HUMAN JUNGLE

As an ecological dominant, the human species has had the opportunity to develop more variety, on the mental plane, than is likely to be found among other species that are held in more strict discipline by nature. The artist who created the *Pogo* cartoon series hit it right on the head: His animal characters do in fact stand for psychological types among people. They can differ from each other as much as do the species in undisturbed wilderness—from alligators to ants and everything in between. Creativity of the human brain and mind is as vast as that.

This explosive diversity of human minds has been harnessed, through much of history, by means of a mental analogue to the beehive and the anthill. The majority of human minds have been trained into conformity with the wider fabric of society not by physical starvation as much as by mental discipline. This has been done with somewhat different patterns of discipline in different societies. The people of Europe, China, India, and so on are carrying the load of past discipline by somewhat different principles. The variation within great civilizations is narrowminded when compared with the lush competition in the biological wilderness. It is broken through at both ends—the primitive and the metropolitan—by drives toward diversity that no church and no dictator can stifle.

On the primitive level, we can still find some of the diversity that the *Pogo* cartoonist was looking at. In real backwoods areas at the outskirts of civilizations, more diversity could escape discipline because people were less dependent upon each other, more in charge of their own lives, individually. At the other end, in the metropolis, the much decried "atomization" tends to disrupt the schemata of conformism. This means, in fact, a reaffirmation of the power of diversity, a gradual return in the direction of the jungle. People may conform in the technical uses of media, such as computers and the Internet, but they will more and more use the leeway of the new media to find their own ways. From the computer viruses to the breakup of Soviet society, people as denizens of a jungle are everywhere awaiting their chance to prevail over any governing systems. Even at the height of technological conformism, there have always been escapees from the system. A few have shown their colors, of genius or of crime, but most have used conformism (on the surface) as a mimicry.

All of this should set warning signs in the path of any attempt at oversimplifying history. On the individual level, there is always the unexpected.

LIMITS TO SCIENCE

In prosaic terms, all the above means, for the social sciences, that they ought to desist a little more often (than they usually do) from believing that a rational explanation or even a somewhat complete description of real-life situations is within their reach. *A fortiori*, the same should hit home even more with the humanists in their search for meaning in the wider fabric of human existence. We cannot explain the whole wide world. Even if we could, we would not be able to read it in a lifetime. For random events, there is no explanation at all. Even explanations of systematic events, where possible, are often too complex to be handled by any one person's intellect.

But we can ask this kind of question: What permanent laws, what regularly recurring features can be traced through all human existence? If this question is asked—not in any inclusive sense, as if our systems could encompass all there is to the matter at hand, but rather in the sense of showing some *leitmotifs*—then this insight into human nature may contribute to a reinterpretation of our own lives, the life of our generation and—although this is the least likely achievement—the direction we may be heading toward. Here we may have a gate of entry into the toolbox of creation.

This phrasing has a twofold bearing on the manner in which the humanities and the social sciences are usually conducted. They all have their counterparts in the arts of real life. The gaps of sheer randomness aside, human reality is far too complex to ever be encompassed by exhaustive scientific analysis. To know your nextdoor neighbor completely, you would have to concentrate on him to the extent that you would not have time to live your own life. In most of the practical affairs that add up to human living, we are forced to rely on something less than the full Socratic *episteme*, the tested knowledge of the scientist. That less ambitious category, the *doxa*, or practically accepted truth, has to be sufficient in many real situations. Lacking proof, our common-sense knowledge merges with the creative constructs of imagination. In war and in love, in the creative and the humdrum, we and others are exchanging a stream of impulses that are normative as often as they are cognitive. Our counteranalysis of the insights, mistakes, purpose, or folly of others (and of ourselves) follows track in a flow that is so rapid and so loaded with detail that only the unchecked, highly intuitive synthetic power of the mind can handle it.

It is in this connection that we come across one of the basic mistakes often made in modern social science, especially in the United States: the belief that understanding human affairs can ever proceed without drawing on the intuitive or artistic powers of empathy. Any analysis of human "behavior" lacking this element will always be thin and skeletal, no matter how outstanding in theory or method. The variability of the human phenomenon is so enormous that it is impossible that one could ever discover plausible hypotheses (other than the most elementary ones) without applying the searching rod of intuition, using the introspective and empathetic powers of the mind, as much as formal logic. The social scientist who does not have within himself something of an artist can only act as technical advisor on a research team, but will never be able, alone by himself, to apply his theories and research findings to real-life situations of normal complexity. The fact that *evidence* must always come from the conscious, discursive procedures we call science must not obscure the fact that intuition is indispensable for formulating meaningful hypotheses in the realm of human affairs.

We need not settle here the question of just what it is that goes into the thing called intuition. Recent findings on extrasensory perception have raised more problems than they have solved, but in principle they have shattered the old construct of the mind as being a closed chamber unable to communicate except through the conventionally known powers of the senses.

The fact that discursive logic and objective analysis can never cover more than part of the human realities we are looking at does not render them useless. They are always better than nothing—provided we are aware of their limitations. They will, among other things, draw the limits within which we as artists (and the professional ones, a fortiori) should limit the exercise of empathy and intuition, whenever the welfare or disaster of the real-life human drama are at stake.

PSEUDOSCIENCE

Scientific analysis applied, even piecemeal, to human affairs can remain useful on the condition that the limitations imposed upon it are understood and the consequences heeded. Theoretical analysis, which claims to know all there is to the matter at hand, can turn into a gimmick that may be worse than no insight at all when the reasoning ultimately is based on assumptions that are arbitrary rather than necessary. Science has no worse adversary than pseudoscience. It is too easy a game to point the finger at Marxism and some of the extravagances of Soviet pseudoscience such as those championed by Marr and Lysenko, or the basically ascientific mentality that used to rage against "revisionism." Revision is too obvious a tool of all bonafide science to be elaborated upon. Closer to home are fallacies often not recognized as such.

Foremost among these is "behaviorism." In its refusal to recognize human consciousness as the axiomatic fact that it is, this school of thought tried to ostracize a problem it saw no means of tackling. The behaviorist belief that there must eventually be a solution to all problems of any human relevance is nothing but an article of faith and an entirely baseless one at that. It has a long history within classical materialism, from Feuerbach all the way to Lenin and Ayn Rand. What we have here is a system of circular logic: When only the senses are recognized as sources of insight, the insights are perforce limited to things determined in terms compatible with the senses.

The mere observation of behavior (rather an empty term in itself) limits inquiry into what is obviously only a part of the subject at hand. From this doctrinal standpoint flow numerous false identifications of problems as well as fallacious conclusions about cause and effect. The natural sciences, which the behaviorists took as their model, do not pretend to operate only on the basis of "observation"—they normally regard theoretical constructs as integral parts of their work. The worst part of the situation is in the fact that many behaviorists actually use their introspection, empathy, and intuition, but in surreptitious ways, to underpin their apparently strictly objective findings. If they were really confining themselves to the cold registration of symptomatic phenomena, the sterility of the approach would become apparent.

More specific examples can be found in fields related to behaviorist psychology. The so-called "psycho-linguists" have thought it feasible to measure shades of meaning (connotative meaning, value overtones) in quantitative terms, without first finding a common yardstick. Instead they used a yardstick that is really of variable size and unstable characteristics. The drive toward premature computerization here called to life a semblance of knowledge where none exists. It is embarrassing to realize that the practical, old-fashioned method of learning a language by listening and imitation yields much more dependable results with much less ado. We should know better than Simplicissimus' German schoolteacher: A mother tongue is learned by suggestion and imitation rather than by conscious grammatical analysis. Foreign idioms learned from quasi-exact dictionaries can bring no more than an illusion of understanding.

The same fallacy as in psycho-linguistics permeates a good deal of contemporary educational psychology for the same reasons. These scholars too often refuse to accept the fact that, in the human personality, the whole was there before the parts and the parts never existed in isolation from the whole. The marasm of "look-say" reading brought this tendency into sharp focus. Subsequently, "hooked on phonics" could be such a remarkable innovation because of the fallacy it attacked.

Such instances are more than just warnings of what might happen. For long years, these schools of thought, and others like them, have played serious havoc with the precarious but real possibilities offered by a less arrogant kind of human insight. The pseudosciences not only emit misleading signals. They also get in the way, hindering others from doing better work. In many quarters, the lesson appears strangely hard to learn.

At the other end of the scale is the intellectual mistake that is common among humanists, particularly in Europe: the belief that intuitive insight can ever become a category of scientific knowledge without ex post facto testing of its results by objective criteria. There are numerous examples already among historians, who all too often accept an intuitive hypothesis merely because it was formulated by an acclaimed historian who is generally believed to have strong intuitive powers. Above all, the specialists on art and literature, engrossed as they are with the human purpose of their subjects in ongoing and contemporary life, often fail to recognize that they are exercising two professions at a time—as scientists and as artists. It is so easy to overlook because both of these activities are perfectly legitimate; only the lack of distinction between them is not legitimate.

So predominant in many cases is the artistic aspect of the humanist— for example, as a literary critic—that it is sometimes forgotten that the humanities do have important subjects that can be handled with strict objectivity. The distinction between science and art in general, not only in the humanities and the social sciences, first makes it possible to draw the limits within which we can legitimately exercise our imagination and creativity in our attempts at understanding and mastering the human realities.

Still another point of confusion is the necessity for humanists, as well as social scientists, to have some of the quality of persons of letters, or of communicators, if they want to be read by a wider public. This again does not have to mean that they must be artistically creative in composing a story. The communicative form can be lent to objective or to imaginative information. We do not have to treat here the wider problem of selective bias (K. Dovring 1959). But the distinction between scientific knowledge and intuitive or empathetic insight must be retained also against the dimension of communicative ability.

For past history, the scientific aspect should be the stronger. The artists get their sway when the past is still alive—as, for instance, with a small selection of great literature that deserves to be interpreted for contemporary people. But most of the affairs of the past have no such direct human message. We want to know them in order to know ourselves— to learn the lessons that Klyuchevsky was talking about.

By unraveling some of the mysteries of our origin, evolution, and development, we hope to gain more insight into the strange commands that force us to live and to act the ways we do. To make the image of ourselves seem alive and appealing, we need no illusions imputed to the past. Those of the present are enough.

SPECIALITIES FIRST

The upshot of the above discussion is a request for austerity in the handling of historical information, as well as in the humanities and social sciences handling contemporary material. They all need "historical method" in the sense of critically peeling away all that is not borne out by reliable evidence and of keeping unsettled hypotheses in a box by themselves. Specifically, it may be necessary to repeat over and over again that single events, because of the random aspect of their occurrence, are not amenable to statements of probability (F. Dovring 1960). Probability being a statistical concept, it can only apply to mass phenomena. On single events, there are only two possibilities: Either we know for sure or we do not know at all. The use of probability on small data series also needs more restraint than is often applied. This extends to the use of preconceived patterns and to patterns derived from a small number of real-world instances. Concepts such as "civilizations," when thought of as holistic entities, will only prejudice our reasoning unless they are backed up by objective evidence. If they are just assumed for want of anything better, they are likely to get in the way and do more harm than good.

An important consequence is the fact that single, complex events often cannot be explained at all. Being unique, they are not covered by piecemeal observation of their several component features in different contexts. In a case at hand, the synergistic effect of all these factors drawn together may well be different from what could have been expected from all the partial analogies on which the explanation is based.

The unique events are background facts, some of which will explain subsequent events even though they themselves cannot be explained. For our purpose, which is to gain a better understanding of the human phenomenon, the obvious starting point must be in seeing whether *any* degree of order, or of general laws, can be discovered. It will be necessary to start with single elements that can be established in masses or in large series. Population change, productive technology, income and wealth and their distribution among the classes of society as well as between societies—such and many analogous facets of existence have to be explored using the tools of specialized disciplines. All along, we must keep under review the leading hypotheses already in use. Some of them may need clarifying, refining, or destroying. Some of the major myths of our time have, in fact, been generated by scientists. In the human universe, as in the physical cosmos, there are but few and simple axioms.

Let us exemplify by citing a few cases of the two opposites—over generalization versus focusing on one factor at a time.

The grandest of all generalizations in modern historiography is obviously the masterpiece by Arnold J. Toynbee, *A Study of History* (1930s). All recorded events since the coming of written documents, in all four corners of the Earth, have here been arranged into a comprehensive system of so many civilizations that have flourished, decayed, and perished. Only the current Western civilization is deemed capable of surviving, at the price of being transformed into a true world civilization.

The last mentioned conclusion is certainly under dispute now, especially so in Asia where serious alternatives are being debated. But the main flaw is not in this kind of prediction, the validity of which would depend on the clarity and necessity of the concept itself. And here Toynbee does, in fact, forsake us because his very concept of civilization includes so much that is subjective, reflecting the author's personality, that the entire exercise is revealed as a vast piece of pseudoscience.

Less subjective is the boldly conceived study of Barrington Moore, Jr., on the origins of dictatorship and democracy (Moore 1966). Critics were generally impressed, more by what the author tried to do than by what he actually accomplished. Through many painstaking analyses, with great erudition, the theme is pursued and yet no very clear general pattern emerges. The author has sought the causes of political outcomes mainly in political events; he did not go far enough into the material preconditions. China's large rivers, and the flood control and irrigation projects they necessitated, parallel earlier stories in Egypt and Babylonia. Geographic determinism goes a long way to set the stage for politics here. By contrast, Britain as an island, with short and small rivers everywhere granting easy ocean access, had quite different geographic preconditions for the evolving of political institutions.

The largest single set of preconditions for political history is in population pressure—demographics in the wide sense. The basic theory has been with us since the time of Malthus (around 1800). His general scheme has often been misunderstood as a generalizing prediction, when in fact it says "if this—then that." Population pressure will be unrelenting and devastating, unless something is done about it, by intent or by default.

Malthusian theory is well illustrated in the long-term history of the Chinese empire and was also noted (with somewhat less rigor in the analysis) by the "Chinese Malthus," Hung Liang Ji, actually, a little earlier than Malthus. Population data as such are skimpy from China (Ho 1959), but the general conditions leading to population pressure are well-known in this case, where the main reward for economic success was the opportunity to have a large family.

Malthusian theory is also well reflected in the recent history of many colonial countries. Peace and orderly administration imposed by the colonial powers gave the dependencies the opportunity to multiply their populations, while other economic opportunity was not forthcoming to match (F. Dovring 1962).

More intriguing are the ways in which population pressure was avoided in some cases. Outstanding are those of ancient Greece and Renaissance Europe. What held down demographic pressure in Greece we can only dimly perceive from the glimpses of cultural history; certainly the vigorous expansion across water also helped. Renaissance Europe is somewhat more clear. It is well established that there was substantial reduction in the population of Western Europe, beginning in the mid-1300s. Causes of this are less well agreed upon, but some of the consequences are clear enough. With fewer people to feed, less grain and more meat was produced—a general improvement in the standard of living. This was important as a backdrop to the popular movements of discontent leading to the Reformation; it also meant more spare resources for cultural diversity, as well as increased demand for spices, which were the original objective in overseas exploration. The vigor of European expansion could hardly have been obtained under grinding population pressure.

Demographics is likely to affect political institutions by way of social stratification. Foremost is the theory of rent, as explained by David Ricardo, and complemented by Karl Marx with the concept of surplus value. Rent is a function of scarcity, explains Ricardo; if there was an abundance of free land (meaning, if population were sparse), then there would be no rent to pay, no basis for distinctions between lords and common people. Settlement history in North America exemplifies this, with the important exception of the South where an overarching "peculiar institution" (slavery) could be upheld by ethnic contrast. The latter points also to the significance of existing political institutions as sometimes perpetuating a scarcity of land that would not be there by itself (F. Dovring 1965).

Population pressure as a direct destroyer of resources is well known from many areas of long standing civilization, but perhaps nowhere as loudly and clearly as in Oceania, from Easter Island to the New Zealand of the Maoris (Flannery 1995).

The fallout of demographics on political institutions by way of class cleavages is well known if not always as clearly perceived. Even the long-term history of the United States displays a trend toward polarization, which at times has been countered by one incidental or another, such as the generous disposal of Federal domain land, the war economy 1915–1918 and 1940–1945, as well as some political solutions to stem the rise of poverty. However, the long-term trend is visible throughout the country's history (F. Dovring 1991).

Political events as such offer less opportunity for systematic analysis, except for the misuse of distorted historical account as a weapon in political struggle. An extreme example is in the story of King Richard III of England. The tradition set in place by his successors, the Tudors (Henry VII and Henry VIII) paints Richard III as a villain with many crimes to his record, finally defeated by the heroic Tudor army in the battle of Bosworth. Peeling away this Tudor tradition has allowed King Richard to come forth in a very different light. Bosworth was not a battle, just a cowardly murder by ambush, and the crimes imputed to King Richard were, in fact, perpetrated by his successors. Laboring under such a political tradition, Shakespeare had, of course, no choice, under Elizabeth I (Tudor), but to dramatize the same legend.

The case of Richard III versus Henry VII is more than just the replacing of one monarch with another. It also meant a watershed in the political and social history of England: The "Tudor enclosures," which dispossessed numerous peasants and enriched some landlords, could not have happened without the Tudor monarchs. Thus a political myth was used to gloss over a crime far greater than the murder of a king.

TO WHAT PURPOSE?

As humans, we have purpose built into our mental fabric. Studying history, we want to discover what it is all about. To date, the only results are speculation. Faith, hope, and love may still be with many of us, but they are not in the province of science.

History has often been read as a morality drama writ large. Like the progress of the Israelites across the desert toward the Promised Land, mankind has been supposed to advance, unknowingly, toward some dramatic conclusion. Recently, some natural scientists have added their versions of "ultimate purpose" in the constructs of "anthropic principles" in the future of mankind. The assumption of an intelligent universe (Hoyle 1983) could be read as a part of the same general framework.

To date, reality instead shows mankind as fumbling and stumbling along on this planet without much understanding of how fragile it is. Destruction of natural resources has been going on for a long time. The pattern of civilization as "future eating" (Flannery 1995) is nothing new, only now the scale is larger.

Science has recently taught us how biological evolution has been disturbed by periods of violent extinction of most living species, in episodes triggered by impact of intruding bodies from outer space, of whatever description. As if such dangers were not enough, people now are causing extinction of living species at rates that may rival some of those natural cataclysms.

In still other ways, industrial civilization plays havoc with fragile environments, oblivious of the cumulation of effects that is likely to lie up ahead. Compared to the now defunct "socialist" regimes of what used to be the Soviet bloc, capitalism may be more efficient at getting more consumption (as conventionally measured) out of the ruined natural resources, but as a brief blip in the history of Earth, it too seems to drift toward physical ruin almost as fast as those defunct regimes.

Man-made deserts and man-made extinctions of biota leave tracks as frightening as those of any natural catastrophies (Bulloch 1989; Asimov and Pohl 1991). *Vestigia terrent.*

If this dance of death at the edge of chaos is going to turn into some kind of happy ending, then no complacency can be tolerated. When cleansed of wishful thinking, very little will remain of conventional American "conservatism" or of the false hopes of traditional progressivism.

As far as studying history for serious purpose, nearly all the real work remains to be done. Only a recognized tabula rasa can serve as background. Know that you know nothing.

Then, maybe, we can begin to build some genuine insight into our real nature and our real prospects.

TEN

What For? Or, What Price Knowledge?

Wissenschaft:
Einem ist sie die hohe, die himmlische Göttin; dem Andern
eine tüchtige Kuh, die ihn mit Butter versorgt
(Science:
To one, she is the exalted and heavenly Goddess; to another
she is a capable cow which keeps him supplied with butter)
Friedrich von Schiller (1759–1805)

Laws of Economics tend to pursue us right into the center of all our search for truth—not the least in economics itself. Science is now such a large and expensive enterprise that in the competition between its various purposes, the question of cost cannot be set apart from that of purpose.

The conflict between the ideal and the useful is not only classical. It is also—at length—self-resolving. The wires are crossing often enough, between pure theory and the tangibly practical. Nuclear physics was discovered out of a drive to know basics, but soon led to the invention of new weapons of war as well as many beneficial applications, in medicine for instance. Laser weapons are another case of the same nexus (Anderberg and Wolbarsht 1992). Telecommunications led from the local community to the creation of a new kind of world community (Clarke 1992).

Heredity was studied by practical breeders of sheep and strawberries and sugar beets, and led to unfathomable insight into the nature of life itself. No one will deny the unity of science or the compatibility of theoretical and practical objectives.

Instead, we are now faced with deep problems as to what we *can* know about and how we may civilize this kind of conflict. And here the classical one tends to sneak in again in a quest to find out what is approximative and what touches the core of reality. Openly accepting approximation, as in probability and fuzzy logic, has great advantages for practical applications, but also tends to postpone any search for absolute truth. Worse, the unrelenting search for meaning in our existence can easily be sidetracked into constructs (as in cosmology, for instance), which are, in fact, a kind of engineering solution writ large.

Concentrating on central theory, on the other hand, will easily lead to extrapolation beyond the reach of observed reality and so lure us into territory where there are no longer any criteria of reality. The heroics of recent cosmology have provoked a pastiche on the old jibe about statistics as lies: We now have (1) speculation, (2) wild speculation, and (3) cosmology. Venturing out to where there are no reality checks, pure theory risks becoming pure confabulation. The analogy with parapsychology is not completely arbitrary. How far out should one push astronomy, really—or psychology? Some recent theorists openly advocate the viewpoint that science is essentially "unnatural" (Wolpert 1992).

Before we go into a discussion of what should have priority for the use of scarce resources in scientific work, we may look briefly at some alternatives as to what is considered "truth."

THE WHOLE TRUTH . . .

The problem begins with mathematics. We already referred to the debate among mathematicians as to whether their formulas reflect what is really out there, or are merely mental constructs to help us in our work. We suggested that there is, in fact, some of each.

The reality out there may be well reflected by conventional mathematics wherever the realities are such that they lend themselves to digital accounting. The whole area of elementary arithmetic just has to be realistic as far as the realities are those of discrete, discretely discernible units. The multiplication table is convincingly explained by a chessboard demonstration.

The other extreme is exemplified by the problem of motion along curved lines. The classical example was formulated by the Eleatic school of Greek thinkers, who maintained that motion is impossible because it would consist of an interminable set of infinitely small parts, and that would take forever. Thus Achilles cannot win his race with a turtle, and the flying arrow will never reach its target.

In modern science, the problem has been "solved" by calculus, which samples the infinitely small parts and then adds up the whole from such a sample of parts. A neat analogy is in motion pictures, which are really a series of still pictures flashed at great speed, leading to an optical illusion of movement. But that is not really what goes on in the ballistics of cannons and rockets.

From my time in the Swedish army, I still vividly recall serving on observation post to watch the arrival of artillery shells as they hit the ground on or near the intended target area. When the shell came from a howitzer, we could actually see it as it flew through the air. A shell of 105 millimeters in diameter looked like a small pea in the sky, revealing its true size only when it hit the ground and exploded. Trying to analyze these visual memories, it is evident that not only were the Eleatics wrong, so is the basis for differential calculus, and maybe even modern physics. The fact of the matter is that no one can tell, at any one instant, where the flying shell is. The instant we could tell, it was no longer there. The shell cannot be said to be in any definable point of space at any definable point in time. The very expression "point in time" is mistaken for "in continuous motion"; time is a flow, a genuine integral—it has no "points." The movement of the flying artillery shell is not composed of small parts. What Heisenberg wrote about locus and momentum must be revised, for there is never any definite locus for an object that is engaged in continuous motion.

To get a practical handle on such a continuum, we resort to approximations, as in calculus. There is no absolute truth here that we can lay our hands on, other than the truth that the continuum, as such, escapes our intellectual tools. To the contrary, computer-based science nowadays increasingly insists on "digitalization" of computation, even where this runs counter to the nature of the things being computed, as the main avenue of progress in science (Negroponte 1995).

The continuity of motion, its integral character, also explains why we can, to some extent, predict future events, at least in the short run, which is encompassed by continuous motion. Knowing the shape of the artillery shell's trajectory, we can predict the time and place of impact—unless something destroys the motion by interception.

The problem of wholeness extends far beyond the simple instance of continuous motion. The class of phenomena we call *synergies* also escapes conventional analysis (or any known analysis, for that matter). So far, any attempt at analyzing synergies is approximate; if it masquerades as exact, it becomes false. The whole truth often eludes us, whether because our knowledge instruments are insufficient or for other reasons. Maybe there are cases in which there could not possibly be any complete access.

In practical reality, we resort to approximations—they "work." How seductive they can be is exemplified by the case of motion pictures. But engineering solutions are not, by definition, "the whole truth." Conflicts such as those between relativity theory and quantum mechanics may reflect different approaches to approximation (von Baeyer 1992).

. . . AND NOTHING BUT . . .

The tendency to exaggerate the validity of recent theory is not limited to the cases of extrapolation into unknown outer reaches of reality. The desire to move toward completeness often refuses to let go even in other cases when there is, in fact, not enough to go by.

One case in point is the theory of evolution. Even Denton (1986), criticizing standard Darwinian theory, utters some regret: If we have to reject Darwinism, then we have no theory of evolution. As if an ill defined theory were better than nothing!

Another case concerns near-death experiences and related phenomena. A recent critic (Blackmore 1993) admits that her alternative theory has weaknesses, but proposes it to be better than no theory at all. Agnosticism is rejected—on what grounds?

What such a standpoint overlooks is the fact that the "better than nothing" theory may be worse than nothing if it blocks further attempts. In such cases, Occam's razor not only does not help, it actually hinders.

At this juncture it may be useful to recall that many dicta of current science are subject to certain doubts, many of which are swept under the rug because they would only leave us with the uncomfortable alternative of admitting ignorance (Collins and Finch 1993). Doubters about current physics theory, for instance, range from Axel Hägerström (1936) to Nancy Cartwright (1983, 1994).

Given the dichotomy of engineering versus absolute science, we are forced to realize that some things are necessary to explore as soon as possible, while some other things can, and indeed may, wait. Searching for direction, we find that in the foreseeable term, research ought to be concentrated where it can be expected, realistically, to yield results that concern our human problems in the here and now.

In some directions we may have to search somewhat less for power over nature and much more for power over ourselves, lest we ruin nature by turning it all into our stifling lighted room (Derber 1996). Throughout, we must insist on the distinction between the approximately true and the absolutely true.

Identifying certain pairs of alternatives may help us organize this discussion. Cosmology versus planetary defense, bioengineering versus ecology, French garden versus jungle (or, suffocation by unity versus a new chaos), virtual reality as master and as servant, media versus content, World War III versus "whither Mankind"—some such paired alternatives (and there are many others that could be proposed) may help in finding the balance of effort when means remain limited. Concentrating on "nothing but the truth" will make us give pause in any quest for "world explanation."

COSMOLOGY AND PLANETARY DEFENSE

It is easy to utter skepticism about the bold cosmology formulated, for instance by Stephen Hawking (1993), as well as about the attendant attempts at "theories about everything" (Penrose 1989; Barrow 1991, 1992; Weinberg 1992; Taylor 1994). The crisis of cosmology has already been articulated elsewhere (Lindley 1993). It is, of course, doubtful whether these extrapolations into events that bear no resemblance to any empirically established facts will eventually prove empirically valid.

It is at least as significant that in conducting our lives, we have no particular use for the notions that the Cosmos was created by the Big Bang, and maybe (but just maybe) will end in a Big Crunch some unspecifiable billions of years from now. Solving human problems on the scale of centuries and maybe millennia, we find that cosmic time becomes practically linear, no matter what kind of large cycle it may belong to.

But there is more to knowledge than what can be used. Astronomy may have been furthered, originally, by the mistaken belief that human fates could be read in the stars as they were then understood or rather misunderstood. Fallacious though it is, astrology in the past has been very productive in promoting scientific inquiry into the real stars. The current search for cosmic explanations may well be a wild-goose chase of dubious value if people are brought to believe in doctrines that are not scientifically necessary. Yet, on the intermediate scale, astronomy has brought a great deal of intellectual satisfaction in clarifying what it is that we all see in the heavens on a starlit night. It was not for nothing that Orwell's imaginary nightmare State in 1984 denied astronomy as well as all other knowledge that did not serve the State.

Recently, however, there has come out of astronomy more than mere intellectual satisfaction. Some of its findings are such that our fate in the foreseeable term may hinge on them. From research in the life forms of past ages (paleontology), we now know that Earth has had some very traumatic episodes of mass extinctions of biological species (Ward 1992) and that these are linked to episodes of collision with bodies from outer space, such as comets and cometary fragments or asteroids and meteorites. Such a collision could happen again within a century or less, so we should worry about them. And we should do more than just worry: It is not excluded that such a cataclysmic event might be averted by some version of "star wars" defense (Steel 1995), using rocket-born explosives to blow the extraterrestrial body out of its threatening course. There are two dilemmas about this. One is that such use of explosives might give a new lease on life to nuclear weapons, after we got rid of them to avert the danger of a "nuclear winter" devastating our planet (Sagan and Turco 1990). The other is in the attempt at cost estimating planetary defense: As an absolute evil, total extinction cannot be compared to any actuarial life table in insurance. A paramount value, such as survival versus extinction, cannot be cost estimated.

Planetary defense evidently concerns much less than the whole gamut of what astronomy deals with. Alien galaxies are still just a matter of curiosity. Even so, it would be foolish for nonastronomers to try to draw the limits for "useful" astronomy. The problems of planetary defense may have remote connections. In some way it could relate to the solar system's trajectory through our galaxy (the Milky Way) because of the clouds of "cosmic dust" and other interstellar debris that may be encountered there. Drawing the line between the useful and the merely interesting must still be left to the astronomers. We can only hope they will heed the necessity to economize with scarce means.

Near-term astronomy may be urgently needed for planetary defense, but as to remote cosmology, even the remotest futurology would make any use of it only very far into the future, *after* a host of human and other near-term problems have been solved (or failed to find any solution). In the meantime, as to the hypothetical Big Crunch, we can only say "que serà, serà."

BIOENGINEERING VERSUS ECOLOGY

Manipulative science threatens to meddle with the innermost secrets of nature. Among other things, someone has envisaged creation of "super-human" beings in the future. "Super" for what purpose? If it is found at all possible, should not such tampering with nature at least wait until we better comprehend what it is that we might be tampering with? The debate is far from closed (Olson 1989; Suzuki and Knudtson 1989; Levine and Suzuki 1993).

The plans for remodeling mankind run parallel with ongoing bioengineering for food farming, promising us "superpigs" and "wonder-corn" (Fox 1992). The matter is not as simple as it may appear on the surface. Other than by higher yielding strains of crops and domestic animals, food farming might better serve the future by supplying more of the crops that serve a vegetarian diet and a less consuming lifestyle. What if pigs become obsolete the moment those "super" ones arrive?

To the contrary, huge amounts of research are needed on near-term, here-and-now problems to avert the impending global catastrophes that now threaten us as a result of mindless, piecemeal technology and "development." Mankind, the ecological dominant, has been attacking nature without understanding what it was attacking, and the cumulative results are now beginning to stare us in the face (Meadows, Meadows, and Randers 1992).

A great deal of meritorious research is being done to support the threatened environment—on biological balance in nature, on macro- and microclimatology, on the oceans as "sinks" for pollutants, and much else. All of this must go on, of course. Some of it is even directed at correcting mistakes in earlier attempts at protecting the environment, as in the defense of the shoreline in the United States (Pilkey and Dixon 1996).

Such research must be complemented by more intensive work on scientific literacy among the voting public, who need antidotes against their wishful thinking masquerading as conservatism. Insight is necessary, lest power over nature lead to blind destruction.

GARDEN VERSUS JUNGLE

Power over nature is easily misused in ways that render nature more uniform, hence less natural. Deprived of the richness of untamed wilderness, nature trades off less quality for the increased quantity of what people demand from it. The extreme instance is the formal French garden of Lenôtre, dating from the grand reign of Louis XIV. "A garden of Lenôtre, correct, ridiculous and charming," wrote Paul Verlaine in *Poèmes Saturniens* (1865). The same nature-changing ideal of style is still repeated piecemeal in strictly trimmed yew and other ornamentals across the length and breadth of American suburbia. The velvet green residential lawn, emulating the monotonous stretches of golf courses, testifies to the will of many people to express their power over nature by stripping nature of some of its engaging diversity. Chemical weed control and other features of brute strength tend to enforce the will toward uniformity, nay, conformity.

The same thing is writ large in the monotonous one-crop fields of modern agriculture. For whatever its rationale, anything worth doing can be done to excess. In recent times, many greedy farmers, in both America and Europe, have sought to maximize their monocultural crops by cutting out windbreaks and hedgerows in order to plant crops on nearly 100 percent of their land—"fencerow to fencerow." Nature responded by erosive dust storms and by the dearth of songbirds as well as of pollinating bumblebees. The lesson was learned in some quarters, at least, and a minimum of variety has been introduced anew by systematic shelter belts—in Denmark, in Nebraska, and in some areas of Africa even. But in temperate farming areas, wildlife remains essentially defeated.

The same battle is engaged in the tropics, where the destruction of rain forests often leads to accelerated dying out of many wildlife species, as well as irreversible loss of topsoil. At the least, farmers in the tropics should be able to use more crop variation than is the norm in temperate regions.

What happens to natural wildlife versus the controlled life of crops and livestock is mirrored in the cultivation of human minds. From grade school to televised "entertainment," teachers and hucksters are doing their level best to cultivate the public toward the "high-productive" uniformity of a farm field. The promise of the Internet to give us again the freedom of untrammeled life may well be smothered by the sheer weight of quantity involved.

VIRTUAL REALITY AS FACT AND AS SEDUCTION

We already pointed to the importance of virtual reality, even in pre-technological human society. From language to money, we are all critically dependent on conventions that exist in our thoughts before they take on the appearance of physical reality and govern a great deal of our daily doing and avoiding.

In all such conventions, virtual reality still had to subordinate under the constraints of the toolbox of creation. Without any such transcendental support, many attempts inevitably fail, as exemplified by the several artificial languages such as Volapük and Esperanto. Computer translation illustrates the same difficulty.

Now, however, computer technology allows some people to enter a world of "virtual reality" without any reality checks. The dangers have been noted repeatedly (Rheingold 1991; Rushkoff 1994; Heim 1993; Larijani 1994).

It has been said time and again that we should use technology, not let it use us. If so, we have to obey the toolbox of creation and not act as if we could create in total independence of it. Part of the difficulty stems from the enterprising people who keep promoting technology whether it serves us or not.

Rather than engaging in free-floating fantasies with no roots in reality, we need to sort out what we *can* believe in.

MEDIA VERSUS CONTENT

The problem got its dominating formulation from Marshall McLuhan, who coined the expression "the medium is the message." As usual in materialist thinking running wild, he viewed his purported reality through the wrong end of the telescope. The notion has spilled over into recent media research, which purports to have found that, in the minds of viewers (listeners, etc.), images conveyed by the media are as real as direct reality (Reeves and Nass 1996).

A medium of communication, such as a telephone receiver or a television set, is no more a message than is any other gadget we can think of. A refrigerator is also a message of sorts—a message of what it can be used for.

But the real messages worth sending and receiving are much less different, over the ages, than ever dreamt of in materialist philosophy. Telecommunications—messages transmitted over long distances, far beyond individual earshot—have existed at least since the Greek victory in the sea battle at Salamis was communicated to the Greek mainland by shouting from island to island across the Aegean Sea. Or ever since the African jungle began listening to the sound of the sorcerers' drums.

The essential commonality of such age-old examples is the existence of a human community across geographic space. This should reduce the modern media to their proper place of acceleration and intensification. The medium is no more of a message than is any gadget offering more efficient service than its technological predecessors. The medium may be *a* message, but it is not *the* message; that has always resided in the contents being communicated. The contents are not a gadget; they are a mental reality distinguished from other mental realities by specific designs and processes. The word (in a book) is still mightier than the sword, or even than the television antenna. Its might depends on what the book contains, not on how it was printed or edited. This bears pondering, indeed, when we try to approach the vast problems of what is worth doing by science, nay, *in* science.

"WHITHER MANKIND"

Reading some of the futurology of advanced cosmologists, we get a vision of superhuman beings eventually becoming masters of the Cosmos, not just our own solar system.

Getting there is, at best, speculative. Even some much less ambitious future vision is, in any event, completely dependent on certain practical solutions being attained in the here-and-now, or at least in the near-term future. Save our earthly environment or else no one will be there to test any of the various future perspectives, on whatever horizon—least of all the very far ones.

The problem is not so much in the future technology of things as it is in the future frame of mind of people, which so far have fallen far short of any constructive anticipation of future conditions. For openers, we should be aware of the fact that mankind is, as yet, far too varied and has too many sharp inner divisions to be able to become a single entity with a single set of long-term objectives. Such an entity cannot be homogenous to any high degree; that would destroy too much of the inner variety that is one of mankind's great assets. Instead, people need to become cosmopolitans, but before that can come about, our contemporary culture needs to cultivate what cosmopolitans we have, despite all the narrowmindedness of our separate lighted rooms. Someone must explain to the well-meaning but single-minded internationalists—to say nothing about the closed-minded nationals—what they are up against in the harnessing of diversity.

More and more among the enlightened scientists now agree that we need a science of mankind. But what should it consist of? Our conclusion is that it demands quite different things from the refinements of natural science.

THE WHOLE AND THE PARTS

In recent times, there has been enormously strong growth in the natural sciences—both their theory and many of their practical applications—and now the power of computerization has come to overshadow the human side of our existence. There is a need to balance the perspective. Physics and astronomy speak to the immensities of raw space and time, with precious few design features. This cannot be compared with the extreme design intensity of human existence. The human psyche contains immensities of design features; it has very little meaning to compare it with physical cosmology. This high design intensity also means extreme variability, as detailed in the essay about history. The difficulty for generalization is evident.

Here we meet again a dichotomy that has shown up here and there in the previous essays—the dichotomy between the comprehension of parts and the fathoming of the whole. Absolute truth seems to beckon us at the extreme ends of the perspective: There may be absolute truth in the counting of individual grains of gravel and in the contemplation of an entire, synergistic whole.

In between, most truths are, at best, approximative, subject to being revised. That is all right as long as we talk about practical (engineering) solutions, but not for ultimate explanations. We have no use for any hypothetical gods, subject to revision under the paradigm of some eventual scientific revolution. The approximate is all we have for solving the problems of the environment. As for ourselves, our human problems, the situation is quite different.

The human soul may seem irrational to those who would rather analyze it to death, climbing down the ladder toward simpler and simpler elements. Such an approach to a science of mankind is doomed to failure, for our soul and mind can only be understood as a complex whole, not as a sum of parts. Bioengineering will not do on this level, neither will any addiction to free-wheeling virtual reality. We need to heed the toolbox of creation, such as it actually is in our case, not as we would want it to appear.

The human phenomenon can only be understood as a system of interacting synergisms, in general more complex than those of ordinary organic chemistry. In short, a science of mankind needs to explore what these synergisms are—which are the irreducible complexities that we must contend with, lest all attempts at shaping the future end up in a mess of logical circles.

Such a search for the real invariants of human nature has hardly begun in earnest. There are bits and pieces, and some of them may lead forward. An interesting one is the suggestion that mankind has a built-in sense for the value of life in general, not just our own: "Biophilia" (Kellert 1995) might be a main key to our future relations with the biosphere, but also to some hitherto unknown synergism in our own nature. A very few other biologists have also begun to feel the need for knowing our connections to the wild (Ward and Kynaston 1995).

There are interesting parallels in the attempts, for instance, in the feminist movement, toward a radical critique of our inherited human society. "Ecofeminism" (Plant 1989; Capra 1996, Ch. 1.) may be as yet untested, but cannot be discarded either until it has been tested by more than the polemics of our current political debate.

The work toward a science of mankind has just barely begun. We cannot afford to pass up any bona fide departures.

References

Agosta, William C. 1992. *Chemical Communication: The Language of Pheromones.* New York: Scientific American Library.

Alkon, Daniel L. 1992. *Memory's Voice: Deciphering the Brain-Mind Code.* New York: HarperCollins Publishers.

Allegre, Claude. 1992. *From Stone to Star: A View of Modern Geology.* Translated by Deborah Kurmes Van Dam. Cambridge, MA: Harvard University Press.

Anderberg, Bengt, and Wolbarsht, Myron L. 1992. *Laser Weapons: The Dawn of a New Military Age.* New York: Plenum Press.

Asimov, Isaac, and Pohl, Frederik. 1991. *Our Angry Earth.* New York: Tom Doherty Associates.

Atwater, P. M. H. 1994. *Beyond the Light: What Isn't Being Said about Near-Death Experience.* New York: Birch Lane Press.

Bache, Christopher M. 1991. *Lifecycles: Reincarnation and the Web of Life.* New York: Paragon House.

Baeyer, Hans Christian von. 1992. *Taming the Atom: The Emergence of the Visible Microworld.* New York: Random House.

Baker, Robert A. 1992. *Hidden Memories: Voices and Visions from Within.* Buffalo, NY: Prometheus Books.

Banerjee, H. N., and Oursler, Will. 1974. *Lives Unlimited: Reincarnation East and West.* Garden City, NY: Doubleday.

Barber, Theodore Xenophon. 1993. *The Human Nature of Birds.* New York: St. Martin's Press.

Barrow, John D. 1988. *The World within the World.* Oxford: Clarendon Press.

———. 1991. *Theories of Everything: The Quest for Ultimate Explanation.* Oxford: Clarendon Press.

———. 1992. *Pi in the Sky: Counting, Thinking, and Being.* Oxford: Clarendon Press.

Berry, Ray, ed. 1992. *The Spiritual Athlete: A Primer for the Inner Life.* Olema, CA: Joshua Press.

Bertalanffy, Ludwig von. 1968. *General Systems Theory: Foundations, Development, Applications.* New York: George Braziller.

———. 1981. *A Systems View of Man.* Edited by Paul LaViolette. Boulder, CO: Westview Press.

Blackmore, Susan. 1993. *Dying to Live: Near-Death Experiences.* Buffalo, NY: Prometheus Books.

Bro, Harmon Hartzell. 1989. *A Seer out of Season: The Life of Edgar Cayce.* New York: New American Library.

Bulloch, David K. 1989. *The Wasted Ocean.* (Introduction by George Reiger.) New York: Lyons and Burford.

Calvin, William H. 1990. *The Cerebral Symphony: Seashore Reflections on the Structure of Consciousness.* New York: Bantam Books.

———, and Ojemann, George A. 1994. *Conversations with Neil's Brain: The Neural Nature of Thought and Language.* Reading, MA: Addison-Wesley Publishing Co.

Candland, Douglas Keith. 1993. *Feral Children and Clever Animals: Reflections on Human Nature.* New York: Oxford University Press.

Capra, Fritjof. 1996. *The Web of Life.* New York: Doubleday.

Carlotto, Mark J. 1991. *The Martian Enigmas: A Closer Look.* Edited by Daniel Drasin. Berkeley, CA: North Atlantic Books.

Carrel, Alexis. 1935. *Man, the Unknown.* New York: Harper & Brothers.

Cartwright, Nancy. 1983. *How the Laws of Physics Lie.* Oxford: Oxford University Press.

———. 1994. *Nature's Capacities and Their Measurements.* Oxford: Oxford University Press.

Casti, John L. 1989. *Paradigms Lost: Images of Man in the Mirror of Science.* New York: William Morrow.

Cavalieri, Paola, and Singer, Peter, eds. 1993. *The Great Ape Project: Equality beyond Humanity.* New York: St. Martin's Press.

Chaisson, Eric. 1988. *Relatively Speaking: Relativity, Black Holes, and the Fate of the Universe.* New York: W. W. Norton.

Cheney, Dorothy, and Seyfarth, Robert M. 1990. *How Monkeys See the World: Inside the Mind of Another Species.* Chicago: University of Chicago Press.

Clarke, Arthur C. 1992. *How the World Was One: Beyond the Global Village.* New York: Bantam Books.

Collins, Harry, and Finch, Trevor. 1993. *The Golem: What Everyone Should Know about Science.* Cambridge, Eng.: Cambridge University Press.

Cone, Joseph. 1991. *Fire under the Sea: The Discovery of the Most Extraordinary Environment on Earth—Volcanic Hot Springs on the Ocean Floor.* New York: William Morrow.

Coveney, Peter, and Highfield, Roger. 1990. *The Arrow of Time: A Voyage through Science to Solve Time's Greatest Mystery.* New York: Fawcett Columbine/Ballantine Books.

————. 1995. *Frontiers of Complexity: The Search for Order in a Chaotic World*. New York: Fawcett Columbine.

Crick, Francis. 1981. *Life Itself: Its Origin and Nature*. New York: Simon and Schuster.

————. 1994. *The Astonishing Hypothesis: The Scientific Search for the Soul*. New York: Charles Scribner's Sons.

Croswell, Ken. 1995. *The Alchemy of the Heavens: Searching for Meaning in the Milky Way*. New York: Anchor Books Doubleday.

Dalrymple, G. Brent. 1991. *The Age of the Earth*. Stanford, CA: Stanford University Press.

Darling, David. 1995. *Soul Search: A Scientist Explores the Afterlife*. New York: Villard Books.

Davies, Paul. 1988. *The Cosmic Blueprint: New Discoveries in Nature's Creative Ability to Order the Universe*. New York: Simon and Schuster.

————. 1992. *The Mind of God: The Scientific Basis for a Rational World*. New York: Simon and Schuster.

————. 1995a. *About Time: The Unfinished Einstein Revolution*. New York: Touchstone Books.

————. 1995b. *Are We Alone? Philosophical Implications of the Discovery of Extraterrestrial Life*. New York: Penguin Books.

————, and Gribbin, John. 1992. *The Matter Myth: Dramatic Discoveries that Challenge our Understanding of Physical Reality*. New York: Simon and Schuster.

Dawkins, Richard. 1989. *The Selfish Gene*. New ed. New York: Oxford University Press.

————. 1995. *River out of Eden: A Darwinian View of Life*. New York: Basic Books.

Dennett, Daniel C. 1991. *Consciousness Explained*. Boston: Little, Brown.

————. 1995. *Darwin's Dangerous Idea: Evolution and the Meaning of Life*. New York: Simon and Schuster.

Denton, Michael. 1986. *Evolution: A Theory in Crisis*. Bethesda, MD: Adler and Adler, 1986. Original edition, 1985.

Derber, Charles. 1996. *The Wilding of America: How Greed and Violence are Eroding Our Nation's Character*. New York: St. Martin's Press.

Dovring, Folke. 1951. "Les méthodes de l'histoire agraire." *Annales: Economies, Sociétés, Civilisations* 6:3: 340–44.

————. 1953. *Agrarhistorien*. Stockholm: Hugo Geber.

————. 1960. *History as a Social Science*. The Hague: Martinus Nijhoff.

————. 1962. "The Opportunity to Multiply." *Journal of Economic History* (December 1961).

————. 1965. "Bondage, Tenure, and Progress." *Comparative Studies in Society and History* 7:3 (April).

————. 1984. *Riches to Rags: The Political Economy of Social Waste*. Cambridge, MA: Schenkman.

————. 1991. *Inequality: The Political Economy of Income Distribution*. New York: Praeger.

————. 1996. *Leninism: Political Economy as Pseudoscience*. Westport, CT: Praeger.

Dovring, Karin. 1951. *Striden Kring Sions Sånger 1–2*. Lund: Gleerupska Universitetsbokhandeln.

————. 1954–1955. "Quantitative Semantics in Eighteenth Century Sweden." *Public Opinion Quarterly* 18:4.

————. 1959. *Road of Propaganda: The Semantics of Biased Communication*. New York: Philosophical Library.

————. 1997. *English as Lingua Franca: Double Talk in Global Persuasion*. Westport, CT: Praeger.

Dressler, Alan. 1995. *Voyage to the Great Attractor: Exploring Intergalactic Space*. New York: Alfred A. Knopf.

Duncan, Lois, and Roll, William. 1995. *Psychic Connections: A Journey into the Mysterious World of Psi*. New York: Delacorte Press.

Earle, Sylvia Alice. 1995. *Sea Change: A Message of the Oceans*. New York: G. P. Putnam's Sons.

Edmondson, Amy C. 1987. *A Fuller Explanation: The Synergetic Geometry of R. Buckminster Fuller*. Boston, Basel, Stuttgart: Birkhäuser.

Eldredge, Niles. 1985. *Time Frames: The Rethinking of Darwinian Evolution and the Theory of Punctuated Equilibria*. New York: Simon and Schuster.

Erickson, Jon. 1992. *Plate Tectonics. Unraveling the Mysteries of the Earth*. New York: Facts on File.

Eve, Raymond A., and Harrold, Francis B. 1991. *The Creationist Movement in Modern America*. New York: Scribner's.

Feynman, Richard P. 1985. *QED: The Strange Theory of Light and Matter*. Princeton, NJ: Princeton University Press.

Flannery, Timothy Fridtjof. 1995. *The Future Eaters: An Ecological History of the Australasian Lands and People*. New York: George Braziller. Original edition, 1994.

Fox, Michael W. 1992. *Superpigs and Wondercorn: The Brave New World of Biotechnology and Where It All May Lead*. New York: Lyons and Burford.

Frisch, Karl von. 1966. *The Dancing Bees: An Account of the Life and Senses of the Honey Bee*. Translated by Dora Isle and Norman Walker. New York: Harcourt Brace Jovanovich.

Fuller, R. Buckminster. 1972. *Intuition*. Garden City, NY: Doubleday.

Gardner, Howard. 1991. *The Unschooled Mind: How Children Think and How Schools Should Teach*. New York: Basic Books.

Gardner, Martin. 1995. *Urantia: The Great Cult Mystery*. Amherst, NY: Prometheus Books.

Goodall, Jane. 1990. *Through a Window: My Thirty Years with the Chimpanzees of Gombe*. Boston: Houghton Mifflin.

Goodman, Jeffrey. 1977. *Psychic Archeology: Time Machine to the Past*. New York: Berkeley Publishing Co.

Gould, James L., and Gould, Carol Grant. 1989. *Sexual Selection*. New York: Scientific American Library.

Gould, Stephen Jay. 1987. *Time's Arrow, Time's Cycle: Myth and Metaphor in the Discovery of Geological Time*. Cambridge, MA: Harvard University Press.

————. 1989. *Wonderful Life: The Burgess Shales and the Nature of History*. New York: Norton and Co.

Gribbin, John. 1990. *Hothouse Earth: The Greenhouse Effect and Gaia*. New York: Grove Weidenfeld.

————. 1992. *Unveiling the Edge of Time: Black Holes, White Holes, Wormholes*. New York: Harmony Books.

Griffin, Donald L. 1984. *Animal Thinking*. Cambridge, MA.: Harvard University Press.

Gurevich, David. 1991. *From Lenin to Lennon: A Memoir of Russia in the Sixties*. San Diego: Harcourt Brace Jovanovich.

Hägerström, Axel. 1936. "Über die Gleichungen der speziellen Relativitätstheorie." In *Adolf Phalén in memoriam*. Edited by I. Hedenius et al. Uppsala and Stockholm: Almquist & Wicksell.

Haken, Hermann. 1978. *Synergetics: Nonequilibrium Phase Transitions and Self-Organization in Physics, Chemistry and Biology*. 2nd enl. ed. Berlin, Heidelberg, New York: Springer Verlag. Original edition, 1977.

Hall, Nina, ed. 1992. *Exploring Chaos: A Guide to the New Science of Disorder*. New York: Norton & Co. Original edition, 1991.

Halpern, Paul. 1992. *Cosmic Wormholes: The Search for Interstellar Shortcuts*. New York: Dutton.

Hawking, Stephen. 1993. *Black Holes and Baby Universes and Other Essays*. New York: Bantam Books.

Heim, Michael. 1993. *The Metaphysics of Virtual Reality*. New York and Oxford: Oxford University Press.

Ho, Ping-ti. 1959. *Studies on the Population of China 1368–1953*. Cambridge, MA: Harvard University Press.

Howells, William. 1993. *Getting Here: The Story of Human Evolution*. Washington, DC: Compass Press.

Hoyle, Fred. 1983. *The Intelligent Universe*. New York: Holt, Rinehart and Winston.

Irwin, H. J. 1989. *An Introduction to Parapsychology*. Jefferson, NC: McFarland & Co.

Jones, Steven. 1994. *The Language of Genes: Solving the Mysteries of our Genetic Past, Present and Future*. New York: Anchor Books.

Jung, Carl Gustav. 1938. *Psychology and Religion*. New Haven, CT: Yale University Press.

Kaufmann, William J., III, and Smarr, Larry L. 1993. *Supercomputing and the Transformation of Science*. New York: Scientific American Library.

Kellert, Stephen R. 1995. *The Value of Life*. Washington, DC: Island Press.

Kornhaber, Arthur. 1988. *Spirit: Mind, Body, and the Will to Existence*. New York: St. Martin's Press.

Kosko, Bart. 1993. *Fuzzy Thinking: The New Science of Fuzzy Logic*. New York: Hyperion.

Krippner, Stanley. 1980. *Human Possibilities: Mind Exploration in the USSR and Eastern Europe*. Garden City, NY: Anchor Press/Doubleday.

Larijani, L. Casey. 1994. *The Virtual Reality Primer*. New York: McGraw-Hill.

Leaky, Richard, and Lewin, Roger. 1992. *Origins Reconsidered: In Search of What Makes Us Human*. New York: Doubleday.

Lear, Jonathan. 1990. *Love and Its Place in Nature: A Philosophical Interpretation of Freudian Psychoanalysis*. New York: Farrar, Straus & Giroux.

Lerner, Eric J. 1991. *The Big Bang Never Happened*. New York: Times Books.

Levine, Joseph, and Suzuki, David. 1993. *The Secret of Life: Redesigning the Living World*. Boston: MGHB Educational Foundation.

Lewin, Roger. 1992. *Complexity: Life at the Edge of Chaos*. New York: Macmillan.

Lightman, Alan. 1991. *Ancient Light: Our Changing View of the Universe*. Cambridge, MA: Harvard University Press.

Lindley, David. 1993. *The End of Physics: The Myth of a Unified Theory*. New York: Basic Books.

Littman, Mark. 1988. *Planets Beyond*. New York: John Wiley and Sons.

Livingston, James D. 1996. *Driving Force: The Natural Magic of Magnets*. Cambridge, MA: Harvard University Press.

Mandelbrot, Benoit. 1983. *The Fractal Geometry of Nature*. New York: W. H. Freeman. Original edition, 1977.

Mann, Alfred K. 1997. *Shadow of a Star: The Neutrino Story of Supernova 1987A*. New York: W. H. Freeman.

Margulis, Lynn, and Sagan, Dorion. 1986. *Microcosmos: Four Billion Years of Evolution from Our Microbial Ancestors*. New York: Summit Books.

———. 1991. *Mystery Dance: The Evolution of Human Sexuality*. New York: Summit Books.

McCrone, John. 1991. *The Ape That Spoke: Language and the Evolution of the Human Mind*. New York: William Morrow.

McGuire, Michael. 1991. *An Eye for Fractals: A Graphic and Photographic Essay*. Redwood City, CA: Addison-Wesley.

McKenna, Terence. 1992. *Food of the Gods: The Search for the Original Tree of Knowledge: A Radical History of Plants, Drugs, and Human Evolution*. New York: Bantam Books.

McNeil, Daniel, and Freiberger, Paul. 1993. *Fuzzy Logic*. New York: Simon and Schuster.

McSween, Harry Y., Jr. 1993. *Stardust to Planets: A Geological Tour of the Solar System*. New York: St. Martin's Press.

Meadows, Donnella H., Meadows, Dennis L., and Randers, Jorgen. 1992. *Beyond the Limits: Confronting Global Collapse, Envisioning a Sustainable Future*. Post Mills, VT: Chelsea Green Publishing Co.

Midgley, Mary. 1992. *Science as Salvation: A Modern Myth and Its Meaning*. London: Routledge.

Monroe, Robert A. 1971. *Journeys out of the Body*. Garden City, NY: Doubleday.

———. 1994. *Ultimate Journey*. New York: Doubleday.

Moody, Raymond A. 1975. *Life after Life: The Investigation of a Phenomenon—Survival of Bodily Death*. New York and Covington, GA: Bantam Books and Mockingbird Books.

———, and Perry, Paul. 1990. *Coming Back: A Psychiatrist Explores Past-Life Journeys*. New York: Bantam Books.

———. 1993. *Reunions: Visionary Encounters with Departed Loved Ones*. New York: Villard Books.

Moore, Barrington, Jr. 1966. *Social Origins of Dictatorship and Democracy*. Boston: Beacon Press.

Morgan, Marlo. 1991. *Mutant Message Down Under*. New York: HarperCollins.

Morse, Melvin, and Perry, Paul. 1990. *Closer to the Light: Learning from Children's Near-Death Experiences*. New York: Villard Books.

————. 1992. *Transformed by the Light: The Powerful Effect of Near-Death Experiences on People's Lives*. New York: Villard Books.

Morton, Eugene S., and Page, Jake. 1992. *Animal Talk: Science and the Voices of Nature*. New York: Random House.

Negroponte, Nicholas. 1995. *Being Digital*. New York: Alfred A. Knopf.

Norman, Diana. 1987. *The Stately Ghosts of England*. Rev. ed. New York: Dorset Press. Original edition, 1963.

Officer, Charles, and Page, Jake. 1993. *Tales of the Earth: Paroxysms and Perturbations of the Blue Planet*. New York and Oxford: Oxford University Press.

Olson, Steve. 1989. *Shaping the Future: Biology and Human Values*. Washington, DC: National Academy Press.

Parker, Barry R. 1988. *Creation: The Story of the Origin and Evolution of the Universe*. New York: Plenum Press.

————. 1990. *Colliding Galaxies: The Universe in Turmoil*. New York: Plenum Press.

————. 1991. *Cosmic Time Travel: A Scientific Odyssey*. New York: Plenum Press.

Penrose, Roger. 1989. *The Emperor's New Mind: Concerning Computers, Minds, and the Laws of Physics*. Oxford: Oxford University Press.

Peterson, Dale, and Goodall, Jane. 1993. *Visions of Caliban: On Chimpanzees and People*. Boston: Houghton Mifflin.

Pilkey, Orrin H., and Dixon, Katharine L. 1996. *The Corps and the Shore*. Washington, DC, and Covelo, CA: Island Press.

Pinker, Steven. 1994. *The Language Instinct*. New York: William Morrow.

Plant, Judith, ed. 1989. *Healing the Wounds. The Promise of Ecofeminism*. Philadelphia, PA, and Santa Cruz, CA: New Society Publishers.

Plotkin, Henry. 1994. *Darwin Machines and the Nature of Knowledge*. Cambridge, MA: Harvard University Press.

Raup, David M. 1991. *Extinction: Bad Genes or Bad Luck?* New York: W. W. Norton.

Raymo, Chet. 1991. *The Virgin and the Mousetrap*. New York: Viking Penguin.

Reanney, Darryl. 1991. *After Death: A New Future for Human Consciousness*. New York: William Morrow.

Reeves, Byron, and Nass, Clifford. 1996. *The Media Equation: How People Treat Computers, Television, and News Media like Real People and Places*. Stanford, CA, and Cambridge, Eng.: Center for the Study of Language and Information, and Cambridge University Press.

Rheingold, Howard. 1991. *Virtual Reality*. New York: Summit Books.

Rifkin, Jeremy, and Howard, Ted. 1980. *Entropy: A New World View*. New York: Viking Press.

Rosenfield, Israel. 1992. *The Strange, Familiar, and Forgotten: An Anatomy of Consciousness*. New York: Alfred A. Knopf.

Rushkoff, Douglas. 1994. *Cyberia: Life in the Trenches of Hyperspace*. San Francisco: Harper.

Sagan, Carl, and Druyan, Ann. 1992. *Shadows of Forgotten Ancestors: A Search for Who We Are*. New York: Random House.

————, and Turco, Richard. 1990. *A Path Where No Man Thought: Nuclear Winter and the End of the Arms Race*. New York: Random House.

Sapp, Jan. 1994. *Evolution by Association: A History of Symbiosis*. Oxford: Oxford University Press.

Savage-Rumbaugh, Sue, and Lewin, Roger. 1994. *Kanzi: The Ape at the Brink of the Human Mind.* New York: John Wiley and Sons.

Seielstad, George A. 1989. *At the Heart of the Web: The Inevitable Genesis of Intelligent Life.* New York: Harcourt Brace Jovanovich.

Shapiro, Robert. 1991. *The Human Blueprint: The Race to Unlock the Secrets of Our Genetic Script.* New York: St. Martin's Press.

Sheehan, William. 1992. *Worlds in the Sky: Planetary Discoveries from Earliest Times through Voyager and Magellan.* Tucson: The University of Arizona Press.

Sheldrake, Rupert. 1991. *The Rebirth of Nature: The Greening of Science and God.* New York: Bantam Books.

Shreeve, James. 1995. *The Neanderthal Enigma: Solving the Mystery of Human Origins.* New York: William Morrow.

Singh, J. A. L., and Zingg, Robert Mowr. 1942. *Wolf Children and Feral Man.* New York: Harper and Brothers.

Skutch, Alexander F. 1996. *The Minds of Birds.* College Station, TX: Texas A&M University Press.

Slouka, Mark. 1995. *War of the Worlds: Cyberspace and the High-Tech Assault on Reality.* New York: Basic Books.

Snow, Chet B. 1989. *Mass Dreams of the Future: Featuring Hypnotic Future-Life Progressions by Helen Wambach.* New York: McGraw-Hill.

Steel, Duncan. 1995. *Rogue Asteroids and Doomsday Comets: The Search for the Million Megaton Menace that Threatens Life on Earth.* New York: John Wiley and Sons.

Steiger, Brad, and Steiger, Sherry Hansen. 1992. *Strange Powers of Pets.* New York: Donald I. Fine.

Stenger, Victor J. 1990. *Physics and Psychics: The Search for a World Beyond the Senses.* Buffalo, NY: Prometheus Books.

Stevenson, Ian. 1974. *Xenoglossy: A Review and Report of a Case.* Charlottesville: University of Virginia Press.

———. 1975. *Cases of the Reincarnation Type.* Charlottesville: University of Virginia Press.

Strum, Shirley C. 1987. *Almost Human: A Journey into the World of Baboons.* New York: Random House.

Suzuki, David, and Knudtson, Peter. 1989. *Genethics: The Clash Between the New Genetics and Human Values.* Cambridge, MA: Harvard University Press.

Talbot, Michael. 1986. *Beyond the Quantum.* New York: Macmillan.

———. 1991. *The Holographic Universe.* New York: HarperCollins.

Taylor, John. 1994. *When the Clock Struck Zero: Science's Ultimate Limits.* New York: St. Martin's Press.

Teller, Edward, Teller, Wendy, and Talley, Wilson. 1991. *Conversations on the Dark Secrets of Physics.* New York: Plenum Press.

Ward, Paul, and Kynaston, Suzanne. 1995. *Wild Bears of the World.* New York: Facts on File.

Ward, Peter Douglas. 1992. *On Methuselah's Trail: Living Fossils and the Great Extinctions.* New York: W. H. Freeman.

Weinberg, Steven. 1992. *Dreams of a Final Theory.* New York: Pantheon Books.

Weiss, Brian L. 1996. *Many Lives, Many Masters.* New York: Time Warner. Original edition, 1988.

Westbroek, Peter. 1991. *Life as Geological Force: Dynamics of the Earth*. New York: W. W. Norton.

Whitton, Joel L., and Fisher, Joe. 1986. *Life between Life: Scientific Explorations into the Void Separating One Incarnation from the Next*. Garden City, NY: Doubleday.

Wilson, Edward O. 1992. *The Diversity of Life*. Cambridge, MA: The Belknap Press of Harvard University Press.

Woese, Carl. 1981. "Archaebacteria." *Scientific American* 244: 98–122.

Wolpert, Lewis. 1992. *The Unnatural Nature of Science*. Cambridge, MA: Harvard University Press.

Zadeh, Lotfi A., and Kaprczyk, Janusz, eds. 1992. *Fuzzy Logic for the Management of Uncertainty*. New York: John Wiley and Sons.

———, and Polak, E. 1969. *System Theory*. New York: McGraw-Hill.

Index